I0014150

JOURNAL OF CYBER SECURITY AND MOBILITY

Volume 4, No. 2–3 (April–July 2015)

Special Issue on
Cybersecurity, Privacy and Trust

Guest Editors:
Samant Khajuria, Camilla Bonde, Roslyn Layton,
Knud Erik Skouby and Lene Tolstrup Sørensen

JOURNAL OF CYBER SECURITY AND MOBILITY

Editors-in-Chief
Ashutosh Dutta, AT&T, USA
Ruby Lee, Princeton University, USA
Neeli R. Prasad, CTIF-USA, Aalborg University, Denmark

Associate Editors
Shweta Jain, York College CUNY, USA
Debdeep Mukhopadhyay, Indian Institute of Technology, Kharagpur, India

Steering Board
H. Vincent Poor, Princeton University, USA
Ramjee Prasad, CTIF, Aalborg University, Denmark
Parag Pruthi, NIKSUN, USA

Advisors
R. Chandramouli, Stevens Institute of Technology, USA
Anand R. Prasad, NEC, Japan
Frank Reichert, Faculty of Engineering & Science University of Agder, Norway
Vimal Solanki, Corporate Strategy & Intel Office, McAfee, Inc, USA

Editorial Board

Sateesh Addepalli, CISCO Systems, USA
Mahbubul Alam, CISCO Systems, USA
Jiang Bian, University of Arkansas for Medical Sciences, USA
Tsunehiko Chiba, Nokia Siemens Networks, Japan
Debabrata Das, IIIT Bangalore, India
Subir Das, Telcordia ATS, USA
Tassos Dimitriou, Athens Institute of Technology, Greece
Pramod Jamkhedkar, Princeton, USA
Eduard Jorswieck, Dresden University of Technology, Germany
LingFei Lai, University of Arkansas at Little Rock, USA
Yingbin Liang, Syracuse University, USA
Fuchun J. Lin, Telcordia, USA
Rafa Marin Lopez, University of Murcia, Spain
Seshadri Mohan, University of Arkansas at Little
Rock, USA
Rasmus Hjorth Nielsen, Aalborg University, Denmark
Yoshihiro Ohba, Toshiba, Japan
Rajarshi Sanyal, Belgacom, Belgium
Andreas U. Schmidt, Novalyst, Germany
Remzi Seker, University of Arkansas at Little Rock, USA
K.P. Subbalakshmi, Stevens Institute of Technology, USA
Reza Tadayoni, Aalborg University, Denmark
Wei Wei, Xi'an University of Technology, China
Hidetoshi Yokota, KDDI Labs, USA
Geir M. Køien, University of Agder, Norway
Nayak Debu, Information and Wireless Security IIT Bombay

Aim
Journal of Cyber Security and Mobility provides an in-depth and holistic view of security and solutions from practical to theoretical aspects. It covers topics that are equally valuable for practitioners as well as those new in the field.

Scope
The journal covers security issues in cyber space and solutions thereof. As cyber space has moved towards the wireless/mobile world, issues in wireless/mobile communications will also be published. The publication will take a holistic view. Some example topics are: security in mobile networks, security and mobility optimization, cyber security, cloud security, Internet of Things (IoT) and machine-to-machine technologies.

Published, sold and distributed by:
River Publishers
Niels Jernes Vej 10
9220 Aalborg Ø
Denmark

River Publishers
Lange Geer 44
2611 PW Delft
The Netherlands

Tel.: +45369953197
www.riverpublishers.com

Journal of Cyber Security and Mobility is published four times a year.
Publication programme, 2015: Volume 4 (4 issues)

ISSN 2245-1439 (Print Version)
ISSN 2245-4578 (Online Version)
ISBN 978-87-93379-42-8 (this issue)

All rights reserved © 2016 River Publishers

No part of this work may be reproduced, stored in a retrieval system, or transmitted in any form
or by any means, electronic, mechanical, photocopying, microfilming, recording or otherwise,
without prior written permission from the Publisher.

JOURNAL OF CYBER SECURITY AND MOBILITY
COMMUNICATIONS

Volume 4, No. 2–3 (April–July 2015)

Editorial Foreword: Special Issue on Cybersecurity, Privacy and Trust

Cybersecurity is a global issue with local impact. When cyber security breaches occur, it impacts on society at all levels: Users lose trust in digital platforms, companies lose money, and public institutions are discredited. Over the last years, cybersecurity as a concept has changed significantly and has become a broad area covering for example crime, espionage, privacy invasions, and technical solutions for prevention of attacks on users, individuals as well as industry/companies. Along with this, privacy has become a central element in the discussion mirroring the fact that users have become much more aware of the consequences of disclosure of private data to the online businesses and market. On the supplier side new as old web- and cloud-based services are aware that they are reliant on the trustworthiness they can provide to the customers. Trust is therefore an underlying element central for any detection, managing or prevention technologies being developed within the field of cybersecurity and privacy. This special issue has seven different contributions addressing aspects of the broad area of cybersecurity.

The first contribution, presents a survey of contemporary botnet detection methods that rely on machine learning for identifying botnet network traffic. The paper titled "On the Use of Machine Learning for identifying Botnet Network traffic" by Matija Stevanovic et al, provides a comprehensive overview on the existing scientific work and contributes to the better understanding of capabilities, limitations and opportunities of using machine learning for identifying botnet traffic. The paper initially presents the background on botnet detection with the focus on network-based detection. In addition to that, the authors also present the principles of using MLAs for identifying botnet network traffic. The paper concludes by outlining possibilities for the future development of machine learning-based botnet detection systems.

The second contribution, titled "Practical Attacks on Security and Privacy Through a Low-Cost Android Device" by Greig Paul et al, addresses the practical risks and vulnerabilities in low-cost Android-based internet tablet designed for the developing world. The paper further discusses the process though which vulnerabilities were identified on the device, as well as the capabilities of each exploit. The authors also discuss the potential attacks which could be carried out against users' ex., Lockscreen Bypass, Theft of User Data, Installation of a keylogger etc.

The third contribution "Comparative Investigation of ARP Poisoning Mitigation Techniques Using Standard Testbed for Wireless Networks" by Goldendeep Kaue et al, discusses and compares various techniques to protect the users from the attacks like man-in-the-middle, denial of service, IP-spoofing etc. Techniques like command prompt, Ettercap, Wireshark and snort are used to identify the attacks. Finally, a comparative analysis is made taking time and scalability into consideration.

The fourth contribution comprises an information security risk assessment of smartphone use in Finland using Bayesian networks. The paper titled "Information Security Risk Assessment of smartphones Using Bayesian Net-works" by Kristian Herland et al. uses a knowledge-based approach method to build a causal Bayesian network model of information security risks and consequences. The papers primary objective is to perform a high-level risk assessment of information security related to smartphone usage. Secondary objective of the paper is to design and implement a practical risk assessment process for eliciting information from multiple experts and consolidation this information into a Bayesian network. The paper concludes by giving a Bayesian network model of information security risks, which can be used for various purposes such as scenario and sensitivity analysis.

The fifth paper titled "Digital Forensic Investigations: Issues of Intan-gibility, Complications and Inconsistencies in Cyber-Crimes" by Ezer Osei Yeboah-Boateng et al. addresses the key challenges such as intangibility, complications and inconsistencies associated with the investigations and presentation of prosecutorial artefacts in digital forensics. The paper unearths the digital truth about malwares and cyber-criminal activities. Furthermore, the study carries out malware analysis, in order to determine the malware activities or operations to comprehend the malware behaviour and analyse the working of the malware codes.

The sixth paper titled "Factors Influencing the Continuance Use of Mobile Social Media: The Effect of Privacy Concerns" by Kwame Simpe Ofori et al. examines privacy concerns in the continuance use of Mobile Social Media. The paper explores the effects of factors such as Perceived Ease of Use, Perceived Usefulness and Perceived Risk and Perceived Enjoyments on Satisfaction and Continuance intention. The papers further analyse a survey data collected from 262 GTUC students using the Partial Least Square approach to Structural Equation Modelling with the use of SmartPLS software.

The seventh contribution discusses various metrics to calculate the trust and evaluation of trust score to determine the trust the user has with the friends in their social network. The paper titled "Confidentiality in Online Networks; A Trust-based Approach" by Vedashree K. Takalkar proposes an architecture that will build trust evaluation system. The paper also discusses the Trust Rule to achieve access control scheme.

Together these seven contributions describe the diversity and range of cybersecurity including trust and privacy. Continued growth in web- and cloud services promoted by the interrelated development of technologies and calls for ever more advanced information societies make these societies increasingly vulnerable to cybercrime and dependant on securing privacy and trust. This implies that more research in these areas is urgently needed as they essential and needed elements in establishing and stabilizing the future services-technology ecosystem.

Samant Khajuria, Camilla Bonde, Roslyn Layton, Knud Erik Skouby
and Lene Tolstrup Sørensen
CMI - Center for Communication, Media and Information technologies,
Aalborg University January 2016

On the Use of Machine Learning for Identifying Botnet Network Traffic

Matija Stevanovic and Jens Myrup Pedersen

Wireless Communication Networks Section, Department of Electronic Systems
Aalborg University, Aalborg, Denmark
Email: {mst; jens}@es.aau.dk

Received 31 August 2015; Accepted 20 November 2015;
Publication 22 January 2016

Abstract

During the last decade significant scientific efforts have been invested in the development of methods that could provide efficient and effective botnet detection. As a result, an array of detection methods based on diverse technical principles and targeting various aspects of botnet phenomena have been defined. As botnets rely on the Internet for both communicating with the attacker as well as for implementing different attack campaigns, network traffic analysis is one of the main means of identifying their existence. In addition to relying on traffic analysis for botnet detection, many contemporary approaches use machine learning techniques for identifying malicious traffic. This paper presents a survey of contemporary botnet detection methods that rely on machine learning for identifying botnet network traffic. The paper provides a comprehensive overview on the existing scientific work thus contributing to the better understanding of capabilities, limitations and opportunities of using machine learning for identifying botnet traffic. Further-more, the paper outlines possibilities for the future development of machine learning-based botnet detection systems.

Keywords: Botnet detection, State of the art, Comparative analysis, Traffic analysis, Machine learning.

Journal of Cyber Security, Vol. 4, 89–120.
doi: 10.13052/jcsm2245-1439.421
© 2016 *River Publishers. All rights reserved.*

1 Introduction

Botnets represent networks of computers compromised with sophisticated bot malware that puts them under the control of a remote attacker [1]. Bot malware provides the attacker with the ability to remotely control behavior of the compromised computers through specially deployed Command and Control (C&C) communication channels. Computers compromised by the bot malware are popularly referred to as bots or zombies, while the attacker is referred to as the botmaster. Controlled and coordinated by the botmaster, botnets represent a collaborative and highly distributed platform for the implementation of a wide range of malicious and illegal activities, such as sending SMAP e-mails, DDoS (Distributed Denial of Service) attacks, Information theft, etc. Due to their malicious potential botnets are often regarded as one of the biggest security threats today [1, 2].

Over the course of the last decade, many botnet detection approaches have been reported in the literature, with various goals, based on diverse technical principles and varying assumptions about bot behavior and the characteristics of botnet network activity [2–4]. As botnets rely on the Internet for both communicating with the attacker as well as for implementing different attack campaigns network traffic analysis is one of the main means of identifying existence of botnets. One of the latest trends in network-based botnet detection is the use of machine learning algorithms (MLAs) for identifying patterns of malicious traffic. The main assumption of machine learning-based methods is that botnets create distinguishable patterns within the network traffic and that these patterns could be efficiently detected using MLAs. This class of detection approaches promises automated detection that is able to generalize knowledge about malicious network traffic from the available observations, thus avoiding pitfalls of signature-based detection approaches that are only able to detect known traffic anomalies. Various detection methods have been developed using an array of MLAs deployed in diverse setups [5–24]. These methods employ diverse principles of traffic analysis targeting various characteristics of botnet network activity. Furthermore, contemporary detection methods have been evaluated using different evaluation methodologies and data sets. The great number of diverse detection solutions introduces the need for a comprehensive approach to summarizing and comparing existing scientific efforts, with a goal of understanding the challenges of this class of detection methods and pinpointing opportunities for the future work.

A number of authors have tried to summarize the field of botnet protection through series of survey papers [1, 2, 25, 26]. Although providing a thorough overview of the field, they only briefly address contemporary detection approaches. In parallel, several authors, such as Feily et al. [25], Bailey et al. [26], Garcia et al. [3], Hyslip et al. [27] and Karim et al. [4] have summarized scientific efforts on botnet detection by proposing novel taxonomies of detection methods and presenting some of the most prominent methods. The authors have acknowledged the potential of machine learning-based approaches in providing efficient and effective detection. Garcia et al. [3] and Karim et al. [4] have provided some of the most comprehensive surveys on existing network-based botnet detection approaches indicating the crucial place of approaches that use MLAs for identifying botnet network activity. Garcia et al. [3] have compared 14 contemporary anomaly-based detection approaches from which 8 were based on MLAs, while Karim et al. [4] analyzed only 3 approaches based on MLAs in more details. Masud et al. [28] and Dua et al. [29] have analyzed the general role of machine learning within modern cyber-security. The authors have outlined the benefits of using machine learning for discovering the existence of the malware on both network and client levels. However, the authors have not provided an overview of the state of the art on botnet detection, leaving the question of current trends within the field of botnet detection unanswered. Finally, several authors have addressed the challenges of using MLAs for network-based detection [30, 31]. Sommer et al. [30] have pointed out some of the challenges and pitfalls of using MLAs for intrusion detection. Although relevant to the realm of botnet detection, claims regarding the usefulness of MLA should be re-evaluated for the botnet detection context. Aviv et al. [31] have presented some of the challenges in experimenting with botnet detection methods that are highly relevant to the use of MLAs for botnet detection.

To the best of our knowledge this paper is the first to provide a comprehensive overview of contemporary detection methods that rely on MLAs for identifying botnet network traffic. The paper has the goal of contributing to a better understanding of capabilities, limitations and opportunities of using machine learning for identifying botnet traffic. The contribution of the paper is three-fold. First, the paper provides a detailed insight on the field by summarizing current scientific efforts. The paper analyzes 20 contemporary detection methods by investigating the principles of traffic analysis used by the detection approaches and how different machine learning techniques are adapted in order to recognize botnet-related traffic patterns. Second, the paper compares the detection methods by outlining their capabilities, limitations and detection performance. Special attention is placed on the practice of experimenting with existing detection approaches and the methodologies of performance evaluation. Third, the paper indicates challenges and limitations of the use of machine learning for identifying botnet traffic and outlines possibilities for future development of machine learning-based botnet detection systems.

The rest of the paper is organized as follows. Section 2 presents the background on botnet detection. The section places special emphasis on botnet detection based on traffic analysis and the use of machine learning for identifying botnet-related traffic. Section 3 presents the analysis principles used in order to evaluate existing detection methods. Section 4 presents the comparative analysis of the state of the art on botnet detection based on machine learning. This section present the most prominent detection approaches by analyzing their characteristics, capabilities and limitations. The discussion of the presented scientific efforts and possibilities for future improvements is presented in Section 5. Finally, Section 6 concludes the paper.

2 Botnet Detection

This section presents the background on botnet detection with the focus on network-based detection. Furthermore, this section presents the principles of using MLAs for identifying botnet network traffic.

Depending of the point of deployment detection approaches can generally be classified as *client-based* or *network-based*. Client-based detection approaches are deployed at the client computer targeting bot malware operating at the compromised machine [28, 32–37]. These methods commonly detect the presence of bot malware by examining different client level forensics, such as: API calls, file changes, application and system logs, active processes, key-logs, usage of the resources, etc. In addition, some client-based detection methods also include the analysis of traffic visible on the computer's network interfaces [22–24]. Network-based detection, on the other hand, provides botnet detection by analyzing network traffic at different points in the network. This class of methods identifies botnets by identifying network traffic produced in different botnet operational phases such as C&C communication, attack phase and propagation [2].

There are several conceptual differences between client- and network-based detection because of which network-based detection is often seen as a more promising solution. Network-based detection is targeting the essential aspects of botnet functioning, i.e. network traffic produced as the result of botnet operation. Network-based approaches assume that in order to implement its malicious functions botnets have to exhibit certain network activity. Botnets could make their operation more stealthy by limiting the intensity of attack campaigns (sending SPAM, launching DDoS attacks, scanning for vulnerabilities, etc.) and by tainting and obfuscating C&C communication. However, this often contradicts the goal of providing the most prompt, powerful and efficient implementation of malicious campaigns. On the other side, attackers invest great efforts in making the presence of bot malware undetectable at compromised machines through a number of client level resilience techniques such as rootkit ability and code obfuscation [38–40]. The attackers also try to deploy a number of network based resilience techniques such as Fast-flux, Domain-flux and encryption but these techniques often introduce additional botnet traits that can be used for detection [17, 41]. Furthermore, as network-based detection is primarily based on passive analysis of network traffic it is more stealthy in its operation and even undetectable to botnet operators in comparison to the client-based detection which could be detected by the malware operating at the compromised machine. Finally, depending of the point of traffic monitoring network-based detection can have a wider scope then the client-level detection systems. When deployed in core and ISP networks network-based detection approaches are able to capture traffic from a larger number of client machines. This provides the ability of capturing additional aspects of botnet phenomena, for instance, group behavior of bots within the same botnet [8, 42], time regularities of bots activity and diurnal propagation characteristics of botnets [43].

2.1 Network-Based Detection

Network-based detection is based on the analysis of network traffic in order to identify presence of compromised computers. Network-based detection is commonly classified based on the principles of functioning as *signature-* or *anomaly-based* methods. Furthermore, methods can be classified as *passive* or *active* depending on the stealthiness of their operation.

The passive detection approaches operate based on observation only thus they do not interfere with botnet operation which makes them stealthy in their operation and undetectable to the attacker. Active detection methods represent more invasive methods that actively disturb botnet operation by interfering with malicious activities or the C&C communication of the bots. Additionally, these techniques often target specific heuristics of the C&C communication or attack campaigns, arguably providing higher accuracy of detection. The majority of contemporary botnet detection approaches are passive while only a few such as BotProbe [44] are active.

Signature-based methods are based on recognizing botnet specific characteristics of traffic, also known as *"signatures"* [45–48]. The signature-based methods rely on a set of predefined rules regarding anomalous traffic and packet level signatures. These approaches commonly performs packet level analysis by using deep packet inspection (DPI) in order to match signatures of malicious payloads. This class of detection techniques covers all three phases of botnet life-cycle and it is able to detect known botnets with bounded number of false positives (FP). The main drawback of signature-based approaches is that they are only able to detect known threats, and that efficient use of these approaches requires constant update of botnet traffic signatures. Additionally, these techniques are liable of various evasion techniques that change signatures of botnet traffic and malicious activities of bots, such as encryption and obfuscation of C&C channel, Fast-flux and DGA techniques, etc.

Anomaly-based detection is a class of detection methods that is devoted to the detection of traffic anomalies associated with botnet operation [5–24, 42, 49–53]. The traffic anomalies that could be used for detection vary from easily detectable as changes in traffic rate and latency, to more finite anomalies in flow patterns. Some of the most prominent anomaly-based approaches detect anomalies in packet payloads [45, 49], DNS traffic [17, 51, 52], botnet group behaviour [8, 42, 43], etc. The anomaly-based detection can be realized using different algorithms ranging from the statistical approaches, machine learning techniques, graph analysis, etc. In contrast to the signature-based detection, the anomaly-based detection is generally able to detect new forms of malicious activity that exhibits anomalous botnet related characteristics. However, one of the main challenges of using anomaly-based detection is the fact that in contrast to signature-based detection these approaches result in false positives. One of the latest and the most promising sub-class of anomaly-based methods are detection methods that rely on machine learning for detection of bot-related traffic patterns. The machine learning is used because it offers the possibility of automated recognition of bot-related traffic patterns. Additionally, machine learning provides the ability of recognizing the patterns of malicious traffic without a prior knowledge about the malicious traffic characteristics, but by inferring knowledge from the available botnet traffic traces.

2.2 Machine Learning for Botnet Detection

The basic assumption behind machine learning-based methods is that botnets produce distinguishable patterns of network activity and that these patterns could be detected by employing some of the MLAs [28, 29].

Machine Learning (ML), is a branch of artificial intelligence, that has the goal of construction and studying of systems that can learn from data [54]. Learning in this context implies ability to recognize complex patterns and make qualified decisions based on previously seen data. The main challenge of machine learning is how to provide generalization of knowledge derived from the limited set of previous experiences, in order to produce a useful decision for new, previously unseen, events. To tackle this problem the field of Machine Learning develops an array of algorithms that discover knowledge from specific data and experience, based on sound statistical and computational principles. Machine learning algorithms can be coarsely classified based on the desired outcome of the algorithm as *supervised* MLAs and *unsupervised* MLAs.

Supervised learning [55] is the class of well-defined machine learning algorithms that generate a function (i.e., model) that maps inputs to desired outputs. These algorithms are trained by examples of inputs and their corresponding outputs, and then they are used to predict output for some future inputs. The supervised MLAs are used for classifying input data into some defined class and for regression that predict continuous valued output. In the context of botnet detection, supervised MLAs are commonly used for implementing network traffic classifiers that are able to classify malicious from non-malicious traffic or identify traffic belonging to different botnets. Some of the most popular supervised MLA used for botnet detection are: SVM (Support Vector Machines), ANN (Artificial Neural Networks), Decision tree classifiers and Bayesian classifier.

Unsupervised learning [56] is the class of machine learning algorithms where training data consists of a set of inputs without any corresponding target output values. The goal of unsupervised learning may be to discover groups of similar examples within the input data, referred to as clustering, to determine the distribution of data within the input space, known as density estimation, or to project the data from a high-dimensional space down to two or three dimensions for the purpose of visualization. In the context of botnet detection, un-supervised MLAs are commonly used for the clustering of bot-related observations. The main characteristic of unsupervised MLAs is that they do not need to be trained beforehand. The most popular unsupervised learning approaches used for botnet detection are: K-means, X-means and Hierarchical clustering.

In both learning scenarios traffic is analyzed from a certain analysis perspective that entails how do traffic instances, that will be classified or clustered by MLAs, look like. For each of traffic instances a set of features is extracted and used within the MLAs to represent them. Choosing the right features representation is one of the most challenging task of practical deployment of MLAs. The chosen features should capture targeted botnet traffic characteristics and pose balanced requirements in terms of feature extraction and selection.

In parallel with the two learning problems outlined here modern machine learning-based approaches commonly implement detection through several phases, using the combination of different MLAs or by deploying MLAs in an adaptive manner. This way more fine grained, flexible, and adaptable detection can be achieved. More details on contemporary detection approaches based on machine learning can be found in Section 4.

3 Principles of the Analysis

This section presents the principles of analyzing contemporary botnet detection approaches that rely on MLAs for identifying botnet network activity. The main goal of the analysis is to provide a review of characteristics and performance of the existing detection methods in order to assess if they can provide accurate, real-time detection that is robust to evasion techniques. The analysis is done by analyzing the characteristics of detection methods, their performance and evasion techniques methods are vulnerable to. The details on the principle of the analysis are presented in the following.

3.1 Characteristics of Detection Methods

The characteristics of detection methods are investigated in order to get the understanding of capabilities and limitations of contemporary detection methods. The analysis of the characteristics is realized by analyzing the following:

- **Point of traffic monitoring** – Point of traffic monitoring infers different capabilities of detection methods. Existing detection approaches monitor traffic at compromised clients, local networks, campus/enterprise networks and in core and ISP networks. The main difference between the points at which methods are implemented is the visible network scope. For instance, a detection system implemented in the core network has the potential of having more comprehensive outlook on the behavior of bots within a certain botnet, than the detection system implemented at a gateway connecting a local network to the Internet. By the same token, the client-based techniques are only able to capture network traffic produced by individual bots. In this analysis we outline points of traffic monitoring assumed by the detection methods.

- **Detection target** – Detection methods commonly detect botnets by identifying bots or C&C infrastructure. However, in this study we also address methods that do not directly detect botnets but provide detection of different network traffic anomalies that commonly characterize botnet operation such as Fast-flux and Domain-flux. These methods are DNS-based detection methods that commonly discover malicious domain names. The findings from DNS-based methods could be used for discovering C&C infrastructure, as well as for discovering potentially compromised clients that try to resolve malicious domain names. In the analysis we outline detection targets of the contemporary detection approaches. Furthermore, we identify all possible detection targets for the DNS-based detection approaches.

- **Botnet type** – Botnets are often coarsely classified based on the employed C&C communication protocol as IRC, HTTP and P2P botnets. Detection methods can cover specific types of botnets or be able to detect botnets independently from the used C&C protocol. Detection approaches that target specific types of botnets are often more efficient than more generic methods. However, these detection techniques are at the same time less flexible to the changes in the communication technology used by botnets. In this analysis we outline the type of botnets targeted by the detection methods.

- **Operational phase** – Detection methods can target different botnet operational phases i.e., the propagation phase, the C&C communication phase or the attack phase [3]. Detection approaches that cover the C&C communication phase can be directed at various communication protocols (IRC, HTTP, P2P), while detection approaches that cover the attack phase can target different attack campaigns (SPAM, DDoS, etc.). Similarly as for the previous characteristic, detection methods that target a specific operational phase can be more efficient than the ones that target more operational phases. Again, these detection techniques are less flexible to the changing nature of the botnet phenomenon. In this analysis we outline operational phases targeted by the detection methods.

- **Communication protocol** – Detection methods analyze traffic at different communication protocols in order to achieve detection. The targeted communication protocols to a large degree depend on the type of botnets and operational phases targeted by the detection approach. Most commonly, detection approaches analyze traffic at transport layer by targeting TCP and UDP traffic and at application layer targeting HTTP, IRC and DNS traffic. In this analysis we outline communication protocols analyzed by the detection methods.

- **MLAs** – The contemporary detection methods rely on different MLAs employed in diverse setups in order to perform detection of malicious network traffic. Both supervised and unsupervised MLAs are used offering different levels of automation and resulting in different types of findings. Furthermore, traffic is analyzed by the MLAs from different perspectives depending on the targeted communication protocols. Finally, in order to capture anomalous characteristics of botnet traffic detection approaches rely on a diverse sets of traffic features extracted for each traffic instance analyzed by the MLAs. In this analysis we outline MLAs and traffic analysis perspective used by the detection methods. Furthermore, we briefly elaborate on the employed feature representation.
- **Real-time operation** – Timely detection is the preferred characteristic of detection systems defined as the ability of operating efficiently and producing the detection results in a "reasonable" time. The timely detection entails a need for a detection method to operate in real-time fashion, thus being capable of processing large quantities of data efficiently. However, it should be noted that the requirements of realtime operation vary depending on the used traffic analysis principles and the goal of detection. In this analysis we outline methods that are advertised as providing real-time detection and detection approaches that have potential of being used in real-time based on their operational efficiency.

3.2 Performance Evaluation

Performance evaluation is realized by analyzing evaluation practices used for experimenting with detection methods, quantitative and qualitative aspects of evaluation data and obtained detection performance.

One of the main prerequisites of reliable performance evaluation is the quality of traffic data sets used for training and testing of the proposed detection approaches [3, 31]. The evaluation data sets should include a substantial amount of traffic for which the *"ground truth"* is known i.e. traffic consisting of elements labeled as malicious and non-malicious. Correctly labeled data sets are one of the main requirements of deterministic evaluation of detection performance. The malicious traffic represent traffic produced by botnets, while non-malicious traffic, often refereed to as the *"background"* traffic originates from benign applications running on *"clean"* computers. The labeled data set is formed either by labeling previously recorded traffic trace or by combining malicious and non malicious data sets. Typical scenarios of obtaining labeled data sets are outlined bellow:

- **Scenario 1** – Labeled traffic is obtained by performing labeling of a network trace using a variety of different labeling practices. The labeling of the traffic can be done by using some of the existing IDS systems [46, 47, 57] and signature-based botnet detection systems [45, 48] or by relying on domain name and IP address blacklists.
- **Scenario 2** – Labeled traffic is obtained by merging a non-malicious traffic trace with a malicious traffic trace captured by Honeypots [58] deployed by researchers themselves or by some third party.
- **Scenario 3** – Labeled traffic is obtained by merging a non-malicious traffic trace with a malicious traffic trace generated within fully controllable testing environments, where researchers have total control on both C&C servers and compromised computers. This scenario requires bot malware source code to be available. Having the source code, experiments can be realized in safe and totally controlled fashion.
- **Scenario 4** – Labeled traffic is obtained by merging a non-malicious traffic trace with a malicious traffic trace generated in semi-controlled testing environments, by purposely infecting computers with a specific bot malware. Compromised computers are allowed to contact the C&C servers in order for bot-related traffic to be recorded. In order to limit any unwanted damage to the third parties on the Internet the traffic produced by infected machines is filtered using different rate and connection limiting techniques as well as matching of the malicious signatures of bot traffic [58].

Non-malicious traffic traces could be obtained in various ways: from self generated traffic using statistical traffic generators to the network traces recorded on LAN, campus/enterprise and in some cases even core and ISP networks. However, it should be noted that for the process of obtaining background traffic the primary concern is to make sure that the traffic traces are benign and "representable" for the particular point of traffic monitoring. While easily achieved on the controlled LAN networks, making sure that traffic obtained from other real-world networks is benign is a much more challenging task. Furthermore, as traffic from one network to another vary, background traffic should match the malicious trace in terms of the point of traffic monitoring and the type of network.

Besides the way evaluation data sets are obtained, the number of distinct bot malware samples used for the evaluation of botnet detection methods is also very important for assessing the generality of the obtained detection performance. Evaluating the detection method using traffic traces from different types of botnets could indicate the ability of the method to cope with new threats. Furthermore, using traffic traces form different botnets for training and testing could give a good indication if a method can generalize well or not.

Understanding performance metrics used for characterizing contemporary detection methods is crucial for assessing the capabilities of approaches. Some of the most frequently used performance metrics are the following:

True positives rate i.e. recall: $TPR = recall = \dfrac{TP}{TP + FN}$

True negative rate: $TNR = \dfrac{TN}{TN + FP}$

False positive rate: $FPR = \dfrac{FP}{FP + TN}$

False negative rate: $FNR = \dfrac{FN}{FN + TP}$

Accuracy: $accuracy = \dfrac{TP + TN}{TP + FP + TN + FN}$

Error rate: $error = \dfrac{FP + FN}{TP + FP + TN + FN}$

Precision: $precision = \dfrac{TP}{TP + FP}$

True positive (*TP*) is the number of positive samples classified as positive, true negative (*TN*) is the number of negative samples classified as negative, false positive (*FP*) is the number of negative samples classified as positive, and false negative (*FN*) is the number of positive samples classified as negative. However, it should be noted that the performance of detection approaches are commonly expressed using only a subset of the presented performance metrics, most commonly TPR and FPR.

3.3 Evasion Tactics

Detection methods should be robust on evasion techniques in such a way that for detection to be evaded botnets should severely limit the efficiency of implementing their malicious agenda. The vulnerability of detection approaches to evasion techniques highly depend on the principles of traffic analysis and the characteristics of botnet traffic targeted by the detection method. Targeting easily changeable botnet characteristics can lead to evasion, which would consequently limit the effectiveness of the detection approach. Following Stinson et al. [59] framework for systematic evaluation of the robustness of botnet detection methods we evaluate the contemporary detection methods against following evasion tactics (ET):

- **ET1** – Evasion of client based detection. Evasion tactic that evades botnet detection at the client machine. This category includes a wide range of techniques, such as evasion by attacking process monitor and evasion by tainting bot malware behavior at the client computer.
- **ET2** – Evasion by traffic encryption. Tactic that performs encryption of the traffic used within the C&C channel.
- **ET3** – Time-based evasion. Evasion techniques that try to avoid bot activity in specific time windows in which detection methods operate, thus restricting the detection methods from catching the right observations.
- **ET4** – Evasion by flow perturbation. This class of evasion techniques changes the patterns of traffic by changing the flow statistics.
- **ET5** – Evasion by performing only a subset of available attacks, thus limiting the available observation for the methods that are targeting the attack phase of botnet operational life-cycle.
- **ET6** – Evasion by restricting the number of attack targets, by targeting clients at the same internal network, thus evading the methods that monitor traffic at network boundaries.
- **ET7** – Evasion of cross-host clustering by employing sophisticated schemes avoiding the group activities of bots within the same administrative domain.
- **ET8** – Evasion by out-of-band coordination of bots, by using Fast-flux and DGA algorithms as a mean of communicating, thus providing a level of privacy and resilience to malicious C&C servers.

The majority of the existing detection methods could be evaded by deploying some of the presented evasion techniques. However, evasion techniques are characterized with implementation costs and performance loss that vary from low to very high [59], thus often causing severe damage to the utility of botnets.

Therefore, the fact that detection system could be evaded does not necessarily mean that the cost of evasion will be justified.

4 State of the Art: The Analysis Outlook

This section analyzes contemporary machine learning-based botnet detection approaches, on the basis of the principles presented in Section 3. The section evaluates 20 contemporary detection methods [5–24]. The majority of the evaluated methods are purely network-based [5–21] while the study also covers several client-level detection approaches that strongly rely on network traffic analysis [22–24].

4.1 Capabilities and Limitations

The characteristics of the analyzed detection approaches are summarized by Table 1 and Table 2, where Table 1 provides an overview of the characteristics of detection methods, while Table 2 summarizes the principles of traffic analysis and MLAs used by the approaches.

Depending on the point of traffic monitoring the majority of detection approaches addressed by this survey monitor traffic at local [9, 10, 12, 21] and possibly campus/enterprise networks [5–8, 11, 20], while others can be implemented in core and ISP networks [8, 13–19]. Finally, some of the client-based techniques that strongly rely on network traffic analysis are also addressed in this survey [22–24].

The detection methods typically contribute to the identification of bots [5–12, 19–21, 23, 24] or malicious C&C servers [13, 18, 22]. DNS-based detection methods [14–17] provide identification of malicious domains that can contribute to the detection of both bots and C&C servers. Based on the identified malicious domains it is possible to identify both bots that try to resolve them as well as the C&C infrastructure behind them.

The majority of the analyzed detection approaches target C&C communication as the main characteristics of botnet operation, while some also include the ability to capture botnet attack campaigns as well [7, 10, 22–24]. The propagation phases is covered by only one detection method [7], most likely as the propagation could be effectively tackled by existing IDS/IPS systems.

Roughly a half of the analyzed detection methods are independent of C&C communication [7, 8, 13–18, 23, 24], while other methods target specific types of botnets, such as IRC-based [5, 6, 12, 22], HTTP-based [21] and P2P-based [9–11, 19, 20] botnets by relying on specific traits of IRC,

Table 1 The characteristics of detection methods

Detection Method	Traffic Monitoring Point	Detection Target	Botnet Type	Operational Phase	Communication Protocol	Real-Time Operation
Livadas et al. [5]	LAN, Campus	Bots	IRC	C&C	TCP, IRC	–
Strayer et al. [6]	LAN, Campus	Bots	IRC	C&C	TCP, IRC	potentially
Gu et al. [7]	LAN, Campus	Bots	Generic	Propagation, C&C, Attack	TCP, UDP	potentially
Choi et al. [8]	Campus, ISP	Bots	Generic	C&C	DNS	advertised
Saad et al. [9]	LAN	Bots	P2P	C&C	TCP, UDP	–
Zhao et al. [10]	LAN	Bots	P2P	C&C, Attack	TCP, UDP	potentially
Zhang et al. [11]	LAN, Campus	Bots	P2P	C&C	TCP, UDP	potentially
Lu et al. [12]	LAN	Bots	IRC	C&C	TCP, UDP	potentially
Bilge et al. [13]	ISP	C&C Servers	Generic	C&C	TCP, UDP	advertised
Bilge et al. [14]	ISP	Bots, C&C Servers	Generic	C&C	DNS	advertised
Antonakakis et al. [15]	ISP	Bots, C&C Servers	Generic	C&C	DNS	potentially
Antonakakis et al. [16]	ISP	Bots, C&C Servers	Generic	C&C	DNS	potentially
Perdisci et al. [17]	ISP	Bots, C&C Servers	Generic	C&C	DNS	potentially
Tegeler et al. [18]	LAN, ISP	C&C servers	Generic	C&C	TCP, UDP	advertised
Zhao et al. [19]	ISP	Bots	P2P	C&C	DNS	–
Zhang et al. [20]	LAN, Campus	Bots	P2P	C&C	TCP, UDP	potentially
Haddadi et al. [21]	LAN	Bots	HTTP	C&C	HTTP	–
Masud et al. [22]	Client	C&C Servers	IRC	C&C, Attack	TCP	–
Shin et al. [23]	Client	Bots	Generic	C&C, Attack	TCP, UDP, DNS	–
Zeng et al. [24]	Client, LAN, Campus	Bots	Generic	C&C, Attack	TCP, UDP	–

Table 2 Traffic analysis perspective and machine-learning algorithms

Detection Method	Analysis Perspective	Supervised/ Unsupervised	MLAs
Livadas et al. [5]	Flow	S	C4.5 Tree, Naive Bayes and Bayesian Network classifiers
Strayer et al. [6]	Flow	S	C4.5 Tree, Naive Bayes and Bayesian Network classifiers
Gu et al. [7]	Client	U	Two level clustering using X-means clustering
Choi et al. [8]	DNS query/response	U	X-means clustering
Saad et al. [9]	Flow	S	SVM, ANN, Nearest Neighbours, Gaussian, and Naive Bayes classifiers
Zhao et al. [10]	Flow	S	Naive Bayes and REPTree (Reduced Error Pruning) Decision Tree
Zhang et al. [11]	Flow	U	Two level clustering using BIRCH algorithm and Hierarchical clustering
Lu et al. [12]	Flow	U	K-means, Un-merged X-means, Merged X-means clustering
Bilge et al. [13]	Flow	S	C4.5, SVM, and Random Forest classifiers
Bilge et al. [14]	DNS query/response	S	C4.5 classifier
Antonakakis et al. [15]	DNS query/response	S, U	X-Means clustering and Decision Tree using Logit-Boost strategy (LAD)
Antonakakis et al. [16]	DNS query/response	S	Random Forest classifier
Perdisci et al. [17]	Clusters of domain names	S	C4.5 classifier
Tegeler et al. [18]	Flow	U	CLUES (CLUstEring based on local Shrinking) algorithm
Zhao et al. [19]	DNS query/response	S	REPTree (Reduced Error Pruning) Decision Tree
Zhang et al. [20]	Flow	U	Two level clustering using K-means algorithm and Hierarchical clustering
Haddadi et al. [21]	Flow	S	C4.5 classifier
Masud et al. [22]	Flow	S	SVM, C4.5, Naive Bayes, Bayes Network, and Boosted decision tree classifiers
Shin et al. [23]	Flow	S	Correlation of the findings of two MLAs: SVM and One Class SVM (OCSVM)
Zeng et al. [24]	Flow	S, U	Correlation of the findings of two MLAs: Hierarchical clustering and SVM

HTTP and P2P C&C channels, respectively. It should be noted that we assume that DNS-based detection methods [14–17] can contribute to the detection of botnets independent of the used C&C communication technology.

The methods analyze different communication protocols in order to perform botnet detection. Based on the analysis TCP, UDP and DNS protocols are the most widely targeted which is reasonable as these traffic protocols cover the majority on botnet network activity. The majority of detection approaches relies on the analysis of TCP and UDP traffic while some more specifically cover IRC [5, 6] and HTTP [21] protocols as they are targeting IRC and HTTP botnets. One approach analyzes all three protocols in order to capture the majority of the botnet network activities [23].

The real-time operation is promised by only a handful of approaches [8, 13, 14, 18]. Some of the contemporary detection approaches show the potential of providing real-time detection as they operate in a time window and they could be periodically re-trained using the new training set or by periodically updating the clusters of the observation [6, 7, 10–12, 15–17, 20]. Finally, some methods such as [14] have proved their ability of real-time operation through a real-world operational deployment.

As illustrated by Table 2, the existing techniques use a variety of machine learning algorithms deployed in diverse setups. In total 15 different MLAs were considered by the analyzed approaches. Supervised and unsupervised MLAs are evenly represented in the analyzed methods. Some of the authors experimented with more than one MLA providing the good insight on how the assumed heuristics hold in different learning scenarios as well as what are the performance of different MLAs [5, 6, 9, 10, 12, 13, 22]. Additionally, some authors used MLAs in more advanced setups, where clustering of observation is realized through two level clustering schemes [7, 11, 20] or where the findings of independent MLAs were correlated in order to pinpoint the malicious traffic pattern [7, 15, 23, 24]. Several authors used the same MLAs within their detection systems [5, 6, 9, 10, 13, 14, 16, 17, 22] indicating some of the well performing clustering and classification algorithms, such as Decision Tree based classifiers and X-means clustering.

The existing methods use several perspectives of traffic analysis. The approaches that analyze TCP and UDP traffic generally analyze it from the perspective of traffic "*flows*". It should be noted that definition of a flow varies from the approach to the approach so some use NetFlow flows [13, 18, 24] while others use a conventional definition of traffic flows where a flow is defined as traffic on a certain 5-tuple i.e. $<ip_{src}, port_{src}, ip_{dst}, port_{dst}, protocol>$. Furthermore, some approaches consider bi-directional flows in order to capture the differences in incoming and outgoing traffic [10]. DNS-based detection approaches commonly analyze DNS traffic from the perspective of DNS query responses (i.e. domain names and their resolving IPs) [8, 14–16, 19], while some analyze it from the perspective of domain clusters [17].

Traffic instances are represented as sets of traffic features in MLAs. As already indicated, feature selection is a challenging task as the feature set should capture targeted characteristics of malicious traffic. The analyzed detection approaches greatly vary in employed feature representation. The TCP/UDP based approaches addressed by the survey use features that are are generally independent from the payload content, relying on the information that can be gathered from packets headers as well as different traffic statistics. Several techniques [7, 12, 21, 22] rely on the content of payloads thus being easily defeated by the encryption or the obfuscation of the packet payload. Furthermore, some approaches rely on IP addresses as features [9, 10] opening the possibility of introducing bias in the evaluation of the detection performance. In the case of DNS analysis approaches typically rely on information extracted from the DNS query responses, such as: lexical domain name features, IP-based features, geo-location features, etc.

4.2 Detection Performance

The analysis of the performance of the methods is illustrated in Table 3, by providing a brief overview of evaluation practice and data sets used within the approaches as well as reported performance for analyzed detection methods. However, it should noted that the results presented in the table represent the bottom range of reported detection performance and that some of the approaches are able to provide better results for specific botnet samples, traffic trace, etc. Additionally, the methods should not be directly compared based on the reported performance alone, as they used different evaluation practices and testing data sets. However, the presented performance can still indicate the overall capabilities of the particular approach in identifying botnet traffic or bots.

Due to the challenges of obtaining training and testing data, the evaluation of the proposed botnet detection systems is one of the most challenging tasks within the development of detection methods [31]. As illustrated in Table 3 the labeled data sets used for development and evaluation of the approaches were obtained through all four scenarios, presented in Section 3. The background data is obtained at the point in the network

Table 3 Evaluation methodology and achieved detection performance

Detection Method	Evaluation Data Sets	Background Data Sets	Number of Botnet Families/ Samples	Detection Performance
Livadas et al. [5]	Scenario 3	Campus	1/1	Flow classification: FPR (10–20%), FNR (30–40%)
Strayer et al. [6]	Scenario 3	Campus	1/1	Flow classification: FPR (<30%), FNR (>2.17%) Bot detection: TPR (90%)
Gu et al. [7]	Scenarios 2, 3, 4	Campus	7/8	Bot detection: TPR (75%–100%), FPR (<1%)
Choi et al. [8]	Scenario 1	Campus, ISP	NA	Domain classification: TPR (>95.4%), FPR (<0.32%)
Saad et al. [9]	Scenario 2	LAN	2/2	Flow classification: TPR (>89%), error (<20%)
Zhao et al. [10]	Scenario 2	LAN	2/2	Flow classification: TPR (>98.1%), FPR (<2.1%)
Zhang et al. [11]	Scenarios 1, 4	LAN, Campus	2/2	Bot detection: TPR (100%), FPR (<0.2%)
Lu et al. [12]	Scenarios 2, 4	ISP	2/2	Flow classification: TPR (>95%)
Bilge et al. [13]	Scenario 1	Campus, ISP	NA	C&C server classification: TPR (>64.3%), FPR (<1%)
Bilge et al. [14]	Scenario 1	ISP	NA	Domain classification: TPR (>98.4%), FPR (<1.1%)
Antonakakis et al. [15]	Scenarios 1, 4	ISP	NA	Domain classification: TPR (>96.8%), FPR (<0.38%)
Antonakakis et al. [16]	Scenario 1	ISP	NA	Domain classification: TPR (>98.1%), FPR (<1.1%)
Perdisci et al. [17]	Scenario 1	ISP	NA	Domain clusters classification: TPR (>99.3%), FPR (<0.15%)
Tegeler et al. [18]	Scenario 4	LAN, ISP	6/188	Bot detection: TPR (49%–100%)
Zhao et al. [19]	Scenario 1	ISP	2/2	Domain classification: TPR (100%), FPR (0.5%)
Zhang et al. [20]	Scenarios 1, 4	LAN, Campus	2/2	Bot detection: TPR (100%), FPR (<0.4%)
Haddadi et al. [21]	Scenarios 1, 3	LAN	3/6	Flow classification: TPR (>84%), FPR (<10%)
Masud et al. [22]	Scenario 3	LAN	2/2	Flow classification: accuracy (>95.2%), FPR (<3.2%)
Shin et al. [23]	Scenario 4	LAN	15/15	Bot detection: TPR (100%), FPR (<0.68%)
Zeng et al. [24]	Scenarios 3, 4	LAN, Campus	5/6	Bot detection: FPR (<0.16%), FNR (<12.5%)

Comment: NA – not available values

corresponding to the monitoring point the methods are developed for, most commonly on campus or LAN networks. A number of approaches obtained evaluation data sets by relying on Scenario 1 [8, 11, 13–17, 19–21]. The majority of these approaches perform DNS traffic analysis so they used domain/IP blacklists and whitelists of popular domains [8, 14–17, 19, 21] for performing labeling while others rely on commercial IDS for doing the labeling [11, 13, 20]. The rest of the approaches relied on other three scenarios of obtaining the evaluation data, where Scenario 2 was used by only 4 approaches, indicating that the researcher needed to run either malware code (Scenario 3) or malware binaries (Scenario 4) in order to obtain malicious network traces.

Furthermore, the malicious traffic samples are usually recorded for a limited number of bot samples. For instance, the performance of only five detection approaches were evaluated on the traffic traces produced by more than 5 bot samples [7, 18, 21, 23, 24], while the maximal number of samples used for evaluation was 188 in case of [18]. The rest of the methods were tested with less than 4 bot malware samples. Finally, the diversity of the used malware samples is poor as the majority of the analyzed approaches rely on less than 3 distinct families of botnets. It should be noted that for DNS-based approaches the number of botnet families that contributed to DNS traffic contained in evaluation data sets is commonly unknown.

The performance reported by the analyzed detection methods indicate a great perspective in identifying botnet traffic and bots using MLAs. Several detection methods indicate TPR of 100% and overall low FPR [11, 19, 20]. Furthermore, a number of approaches is characterized with a FPR less than 1%. These results indicate the possibility of using some of the approaches in real-world operational networks.

4.3 Vulnerability to Evasion Techniques

Table 4 illustrates how different approaches cope against evasion techniques presented in Section 3, by indicating the strength of the indication (SF – strong factor and WF – weak factor) of methods being evaded by them. However, it should be noted that the indications given by Table 4 should be used more as guidelines than as precise measures.

As illustrated in Table 4, the proposed approaches are more or less vulnerable on different evasion techniques. Generally, the majority of the analyzed methods are resistant to evasion by encryption of botnet traffic. Only four approaches [7, 12, 21, 22] that rely on features extracted from packet payload are vulnerable on this evasion strategy. However, the majority

Table 4 An overview of evasion tactics for detection methods

Detection Method	Evasion Tactics							
	ET1	ET2	ET3	ET4	ET5	ET6	ET7	ET8
Livadas et al. [5]	–	–	SF	SF	–	–	–	–
Strayer et al. [6]	–	–	SF	SF	–	–	SF	–
Gu et al. [7]	–	WF	WF	SF	SF	WF	SF	–
Choi et al. [8]	–	–	WF	–	–	–	SF	SF
Saad et al. [9]	–	–	SF	SF	–	–	–	SF
Zhao et al. [10]	–	–	SF	SF	SF	WF	–	SF
Zhang et al. [11]	–	–	SF	SF	–	WF	–	SF
Lu et al. [12]	–	SF	SF	SF	–	–	–	SF
Bilge et al. [13]	–	–	SF	SF	–	–	SF	SF
Bilge et al. [14]	–	–	SF	–	–	–	SF	–
Antonakakis et al. [15]	–	–	WF	–	–	–	SF	–
Antonakakis et al. [16]	–	–	WF	–	–	–	SF	–
Perdisci et al. [17]	–	–	SF	–	–	–	SF	–
Tegeler et al. [18]	–	–	SF	SF	–	WF	SF	WF
Zhao et al. [19]	–	–	SF	–	–	–	SF	WF
Zhang et al. [20]	–	–	SF	SF	–	–	–	SF
Haddadi et al. [21]	–	SF	SF	SF	–	–	–	SF
Masud et al. [22]	SF	SF	SF	SF	–	–	–	–
Shin et al. [23]	SF	–	SF	SF	WF	–	–	WF
Zeng et al. [24]	SF	–	SF	SF	–	WF	SF	–

Comment: SF – strong factor, WF – weak factor

of the techniques are vulnerable on evasion by flow perturbation, due to the fact that they analyze traffic on the flow level. Furthermore, all the techniques are more or less vulnerable on time-based evasion, especially the ones that promise the real-time operation. Finally, the analyzed client-based [22–24] techniques are vulnerable on evasion techniques that target the monitoring of the internals of a host computer.

This study does not address the complexities of evasion techniques and their effect on the overall utility of the botnet. The future work could be directed at more thorough analysis of evasion techniques and vulnerability of modern detection systems to them, covering effect of evasion techniques to both detection systems and overall utility of the botnet.

5 Discussion

In this section we elaborate on the findings of the analysis presented in the previous section outlining several challenges of using MLAs for identifying botnet network activity.

5.1 Principles of Traffic Analysis

As shown in Section 3 detection methods target botnets from different monitoring points where the majority of the approaches are implemented at local and campus/enterprise networks. This can be explained due to several reasons. First, MLAs are data-driven and they depend on available evaluation data sets that are mostly formed by capturing botnet network activity at local networks. Second, the analysis of traffic in core and ISP networks assumes the ability of processing a substantial amount of traffic which can be a challenge for some of the approaches. However, it should be noted that depending on the monitoring point different portions of a botnet can be seen and consequently different malicious traffic characteristics can be targeted. Therefore, detection methods can often be seen as complementary solutions to the botnet detection problem rather than competing solutions.

The analyzed detection methods can either target all botnets types or a specific type based on their C&C communication technology. Furthermore, the approaches could capture different botnet operational phases. The idea of targeting specific type of botnets is to achieve better detection performances which has proved to be true by a number of studies [11, 20]. However, these more specific detection approaches should be used in combination with other detection approaches in order to provide effective and hard to evade detection.

The analyzed detection approaches most commonly target TCP, UDP and DNS protocols using different traffic detection perspectives. We would like to stress the importance of the analysis perspective as choosing the analysis perspective is important to the practical use of machine-learning based approaches. The used analysis perspective should encompass the nature of the targeted phenomena and should carry the context that would be understandable to the operator of the system. A suitable example would be DNS traffic analysis, where DNS traffic could be analyzed from the perspective of either domain names or more complex domains-to-IPs mappings. The former implies that for each domain name a number of features is extracted and MLAs is used to identify if domain is malicious or not. The latter extracts domains-to-IPs mappings i.e. mappings between queried domain names and resolving IP addresses and determines if the mappings are malicious or not. The difference between the two scenarios is significant. Analyzing DNS from the perspective of domain names would segment traffic into much larger sets of instances than in the case of domains-to-IPs mappings. Furthermore, in the first case the result of detection would bring only limited information to the operator, i.e. only detection result. Analyzing DNS traffic from the perspectives of domains-to-IPs mappings would, on the other hand, yield more descriptive detection results as the identified mappings bring more information to the system operator then the simple classification of domain names. Furthermore, domains-to-IPs perspective encompasses the characteristics of IP- and Domain-flux strategies that are often used by the cyber criminals.

As presented in Table 2 different MLAs have been used as a tool for identifying botnets. Some of the most popular MLAs are decision tree classifiers (C4.5, Random Forests, REPTree) for classification and Hierarchical clustering and X-means clustering for grouping traffic observations. This does not come as surprise as these algorithms have showed their capabilities in network traffic analysis and classification of Internet traffic over the last decade [60]. Furthermore, as indicated in the previous section detection approaches use diverse traffic features for representing traffic instances within their detection algorithms. However, using features that would introduce bias in the detection such as IP addresses should be avoided. Also the use of features based on packet payloads should be avoided due to possible evasion by encryption and the violation of the privacy of end-users.

5.2 Evaluation Challenge

The main challenge in evaluating the analyzed detection methods is obtaining reliable evaluation data sets that would successfully capture a substantial amount of both malicious and non-malicious traffic instances. As illustrated in the previous section, existing studies have used different strategies for obtaining the ground truth on botnet network activity, using all scenarios outlined in the Section 3. However, each of the scenarios comes with drawbacks that should be thoroughly understood. The main drawback of Scenario 1 is inherited from the imperfections of data set labeling techniques. Labeling by relying on domain and IP blacklists and signature-based IDSs has its drawbacks that could lead to unreliable ground truth [61–64]. Although many authors [65] have tried to solve this problem proposing different strategies of eliminating false positives in the process of labeling, the problem remains largely unsolved. The scenarios that rely on merging the malicious and non-malicious traffic traces suffer from the pitfall of artificially merging diverse traces. The technique of merging should not introduce additional traffic anomalies that would lead to a biased detection. Furthermore, the non-malicious background traffic used for forming the evaluation data sets should be obtained at the point of traffic monitoring corresponding to the monitoring point at which the botnet trace was recorded.

Based on the analysis the majority of the detection approaches were evaluated using traffic from a modest number of botnets. This could be explained due to many legal, ethical and practical limitations of obtaining the botnet traffic traces. However, the small number of used bot samples indicates the need for more thorough testing where more comprehensive set of malware samples would be used, in order to prove that the detection system is able to generalize inferred botnet knowledge to previously unseen botnets.

5.3 Cost of Errors

The evaluated botnet detection methods report overall promising performances of identifying malicious traffic. However, the majority of the approaches are characterized with a substantial number of FP and FN. We would like to elaborate on the cost of these errors by following Sommer et al. [30] analysis of the cost of errors on intrusion detection use case. As in the case of any other anomaly detection system botnet detection systems are sensitive to number of FP. The high number of FP positives can easily deem the detection method unusable from the perspective of the use in operational networks. Furthermore, depending on the use case botnet detection can have a high cost of FN as any compromised machine within the operational network could cause a lot of technical and financial damage. Therefore, an optimal detection approach would have FP close to zero and minimized number of FN. However, based on the analysis presented in Section 4 only a few detection methods could potentially fulfill this requirement opening the space for further improvements.

5.4 Opportunities for Future Work

Future work should be devoted to the development of detection methods that would find their use within operational networks. These methods should rely on the principles of network traffic analysis that would encompass targeted botnet network characteristics and would carry the context that is understandable to the operator of the system. Furthermore, one of the important goals of future detection systems is to operate in real-time thus facilitating timely detection. The future methods should be evaluated using an extensive set of network traces originating from different types of botnets. Finally, special attention should be placed on minimizing the number of errors in identifying botnet network traffic so the proposed methods would performance-wise be suitable for being used in operational networks.

6 Conclusion

The use of machine learning algorithms (MLAs) for identifying botnet network traffic has been the subject of interest within the research community during the last decade resulting in numerous detection methods. The contemporary detection methods are based on different principles of traffic analysis, they target diverse traits of botnet network activity using a variety of machine learning algorithms and they consequently provide varying performance of detection. This paper outlines the opportunities and challenges of using MLAs for identifying botnet network activity and presents the review of the most prominent contemporary botnet detection methods based on MLAs. The presented study covers 20 detection methods, proposed over the last decade. The methods have been analyzed by investigating principles of their operation, the used evaluation procedures, obtained performance and vulnerabilities on evasion techniques. The analysis indicates a great potential of this class of approaches to be used for identifying botnet network traffic. However, the study also indicates some of the challenges of using MLAs in the context of network-based botnet detection that should be thoroughly understood in order for this class of detection methods to be effectively used.

References

[1] Hogben, G. (ed.), "Botnets: Detection, measurement, disinfection and defence," ENISA, Tech. Rep., 2011.

[2] S. S. Silva, R. M. Silva, R. C. Pinto, and R. M. Salles, "Botnets: A survey," *Computer Networks*, vol. 1, no. 0, pp. –, 2012.

[3] S. García, A. Zunino, and M. Campo, "Survey on network-based botnet detection methods," *Security and Communication Networks*, vol. 7, no. 5, pp. 878–903, 2014.

[4] A. Karim, R. B. Salleh, M. Shiraz, S. A. A. Shah, I. Awan, and N. B. Anuar, "Botnet detection techniques: Review, future trends, and issues," *Journal of Zhejiang University SCIENCE C*, vol. 15, no. 11, pp. 943–983, 2014.

[5] C. Livadas, R. Walsh, D. Lapsley, and W. T. Strayer, "Using machine learning techniques to identify botnet traffic," in *Proceedings of*

2006 31st IEEE Conference on Local Computer Networks, Nov. 2006, pp. 967–974.

[6] W. T. Strayer, D. Lapsely, R. Walsh, and C. Livadas, "Botnet detection based on network behaviour," in *Botnet Detection*, ser. Advances in Information Security. Springer, 2008, vol. 36, pp. 1–24.

[7] G. Gu, R. Perdisci, J. Zhang, and W. Lee, "Botminer: Clustering analysis of network traffic for protocol- and structure-independent botnet detection," in *Proceedings of the 17th conference on Security symposium*, 2008, pp. 139–154.

[8] H. Choi and H. Lee, "Identifying botnets by capturing group activities in dns traffic," *Computer Networks*, vol. 56, no. 1, pp. 20–33, 2012.

[9] S. Saad, I. Traore, A. Ghorbani, B. Sayed, D. Zhao, W. Lu, J. Felix, and P. Hakimian, "Detecting p2p botnets through network behavior analysis and machine learning," in *2011 Ninth Annual International Conference on Privacy, Security and Trust (PST)*, July 2011, pp. 174–180.

[10] D. Zhao, I. Traore, B. Sayed, W. Lu, S. Saad, A. Ghorbani, and D. Garant, "Botnet detection based on traffic behavior analysis and flow intervals," *Computers & Security*, vol. 39, pp. 2–16, 2013.

[11] J. Zhang, R. Perdisci, W. Lee, U. Sarfraz, and X. Luo, "Detecting stealthy P2P botnets using statistical traffic fingerprints," in *2011 IEEE/IFIP 41st International Conference on Dependable Systems and Networks (DSN), Hong Kong*. IEEE/IFIP, Jun. 2011, pp. 121–132.

[12] W. Lu, G. Rammidi, and A. A. Ghorbani, "Clustering botnet communication traffic based on n-gram feature selection," *Computer Communications*, vol. 34, pp. 502–514, 2011.

[13] L. Bilge, D. Balzarotti, W. Robertson, E. Kirda, and C. Kruegel, "Disclosure: Detecting botnet command and control servers through large-scale netflow analysis," in *Proceedings of the 28th Annual Computer Security Applications Conference*, ser. ACSAC '12. ACM, 2012, pp. 129–138.

[14] L. Bilge, S. Sen, D. Balzarotti, E. Kirda, and C. Kruegel, "Exposure: A passive dns analysis service to detect and report malicious domains," *ACM Transactions on Information and System Security (TISSEC)*, vol. 16, no. 4, p. 14, 2014.

[15] M. Antonakakis, R. Perdisci, D. Dagon, W. Lee, and N. Feamster, "Building a dynamic reputation system for dns," in *Proceedings of the 19th USENIX conference on Security*, ser. USENIX Security'10. Berkeley, CA, USA: USENIX Association, 2010, pp. 18–18.

[16] M. Antonakakis, R. Perdisci, W. Lee, N. Vasiloglou II, and D. Dagon, "Detecting malware domains at the upper dns hierarchy." in *USENIX Security Symposium*, 2011, p. 16.

[17] R. Perdisci, I. Corona, and G. Giacinto, "Early detection of malicious flux networks via large-scale passive dns traffic analysis," *IEEE Transactions on Dependable and Secure Computing*, vol. 9, no. 5, pp. 714–726, 2012.

[18] F. Tegeler, X. Fu, G. Vigna, and C. Kruegel, "Botfinder: Finding bots in network traffic without deep packet inspection," in *Proceedings of the 8th international conference on Emerging networking experiments and technologies*. ACM, 2012, pp. 349–360.

[19] D. Zhao and I. Traore, "P2p botnet detection through malicious fast flux network identification," in *2012 IEEE Seventh International Conference on P2P, Parallel, Grid, Cloud and Internet Computing (3PGCIC)*, 2012, pp. 170–175.

[20] J. Zhang, R. Perdisci, W. Lee, X. Luo, and U. Sarfraz, "Building a scalable system for stealthy p2p-botnet detection," *IEEE Transactions on Information Forensics and Security*, vol. 9, no. 1, pp. 27–38, 2014.

[21] F. Haddadi, D. Runkel, A. N. Zincir-Heywood, and M. I. Heywood, "On botnet behaviour analysis using gp and c4.5," in *Proceedings of the 2014 conference companion on Genetic and evolutionary computation companion*. ACM, 2014, pp. 1253–1260.

[22] M. Masud, T. Al-khateeb, L. Khan, B. Thuraisingham, and K. Hamlen, "Flow-based identification of botnet traffic by mining multiple log files," in *First International Conference on Distributed Framework and Applications, 2008. DFmA 2008*, Oct. 2008, pp. 200–206.

[23] S. Shin, Z. Xu, and G. Gu, "EFFORT: Efficient and Effective Bot Malware Detection," in *Proceedings of the 31th Annual IEEE Conference on Computer Communications (INFOCOM'12) Mini-Conference*, March 2012, pp. 71–80.

[24] Y. Zeng, X. Hu, and K. Shin, "Detection of botnets using combined host- and network-level information," in *2010 IEEE/IFIP International Conference on Dependable Systems and Networks (DSN)*, 28 2010-July 1 2010, pp. 291–300.

[25] M. Feily and Shahrestani, "A survey of botnet and botnet detection," *Third International Conference on Emerging Security Information, Systems and Technologies, 2009. SECURWARE '09*, pp. 268–273, 2009.

[26] M. Bailey, E. Cooke, F. Jahanian, Y. Xu, and M. Karir, "A survey of botnet technology and defenses," in *Conference For Homeland Security,*

2009. CATCH '09. Cybersecurity Applications Technology, March 2009, pp. 299–304.

[27] T. Hyslip and J. Pittman, "A survey of botnet detection techniques by command and control infrastructure," *Journal of Digital Forensics, Security and Law*, vol. 10, no. 1, pp. 7–26, 2015.

[28] M. Masud, L. Khan, and B. Thuraisingham, *Data Mining Tools for Malware Detection*. Taylor & Francis Group, 2011.

[29] S. Dua and X. Du, *Data mining and machine learning in cybersecurity*. Boca Raton, FL: CRC Press. xxii, 234 p. $ 89.95, 2011.

[30] R. Sommer and V. Paxson, "Outside the closed world: On using machine learning for network intrusion detection," in *2010 IEEE Symposium on Security and Privacy (SP)*, IEEE, 2010, pp. 305–316.

[31] A. J. Aviv and A. Haeberlen, "Challenges in experimenting with botnet detection systems," in *Proceedings of the 4th conference on Cyber security experimentation and test*, ser. CSET'11. Berkeley, CA, USA: USENIX Association, 2011, pp. 6–6.

[32] M. Bailey, J. Oberheide, J. Andersen, Z. M. Mao, F. Jahanian, and J. Nazario, "Automated classification and analysis of internet malware." in *RAID*, ser. Lecture Notes in Computer Science, C. KrÃijgel, R. Lippmann, and A. Clark, Eds., vol. 4637. Springer, 2007, pp. 178–197.

[33] E. Stinson and J. C. Mitchell, "Characterizing bots' remote control behavior," in *Botnet Detection*, ser. Advances in Information Security, W. Lee, C. Wang, and D. Dagon, Eds. Springer, 2008, vol. 36, pp. 45–64.

[34] L. Liu, S. Chen, G. Yan, and Z. Zhang, "Bottracer: Execution-based bot-like malware detection," in *Proceedings of the 11th international conference on Information Security*, ser. ISC '08. Berlin, Heidelberg: Springer-Verlag, 2008, pp. 97–113.

[35] C. Kolbitsch, P. M. Comparetti, C. Kruegel, E. Kirda, X. Zhou, and X. Wang, "Effective and efficient malware detection at the end host," in *Proceedings of the 18th conference on USENIX security symposium*, ser. SSYM'09. Berkeley, CA, USA: USENIX Association, 2009, pp. 351–366.

[36] U. Bayer, P. M. Comparetti, C. Hlauschek, C. KrÃijgel, and E. Kirda, "Scalable, behavior-based malware clustering." in *NDSS*. The Internet Society, 2009, pp. 5–5.

[37] Y. Park, Q. Zhang, D. Reeves, and V. Mulukutla, "Antibot: Clustering common semantic patterns for bot detection," in *2010 IEEE 34th*

Annual Proceedings on Computer Software and Applications Conference (COMPSAC), July 2010, pp. 262–272.

[38] M. Egele, T. Scholte, E. Kirda, and C. Kruegel, "A survey on automated dynamic malware-analysis techniques and tools," *ACM Comput. Surv.*, vol. 44, no. 2, pp. 6:1–6:42, Mar. 2008.

[39] I. You and K. Yim, "Malware obfuscation techniques: A brief survey," in *2010 International Conference on Broadband, Wireless Computing, Communication and Applications (BWCCA)*, Nov. 2010, pp. 297–300.

[40] J. Marpaung, M. Sain, and H.-J. Lee, "Survey on malware evasion techniques: State of the art and challenges," in *2012 14th International Conference on Advanced Communication Technology (ICACT)*, Feb. 2012, pp. 744–749.

[41] Damballa, "A new iteration of the tdss/tdl4 malware using dga-based command-and-control," Damballa, Tech. Rep., 2012.

[42] A. Karasaridis, B. Rexroad, and D. Hoeflin, "Wide-scale botnet detection and characterization," in *Proceedings of the first conference on First Workshop on Hot Topics in Understanding Botnets*, ser. HotBots'07. Berkeley, CA, USA: USENIX Association, 2007, pp. 7–7.

[43] D. Dagon, C. Zou, and W. Lee, "Modeling botnet propagation using time zones," in *Proceedings of the 13 th Network and Distributed System Security Symposium NDSS*, 2006, pp. 7–7.

[44] G. Gu, V. Yegneswaran, P. Porras, J. Stoll, and W. Lee, "Active botnet probing to identify obscure command and control channels," in *Proceedings of the 2009 Annual Computer Security Applications Conference*, ser. ACSAC '09. Washington, DC, USA: IEEE Computer Society, 2009, pp. 241–253.

[45] G. Gu, P. Porras, V. Yegneswaran, M. Fong, and W. Lee, "BotHunter: Detecting malware infection through IDS-driven dialog correlation," in *Proceedings of the 16th USENIX Security Symposium, San Jose, California.* USENIX Association, Jul. 2007, pp. 167–182.

[46] V. Paxson, "Bro: A system for detecting network intruders in real-time," *Computer Networks*, vol. 31, no. 23ÃćâĆňâĂĲ24, pp. 2435–2463, 1999.

[47] M. Roesch, "Snort - lightweight intrusion detection for networks," in *Proceedings of the 13th USENIX conference on System administration*, ser. LISA '99. Berkeley, CA, USA: USENIX Association, 1999, pp. 229–238.

[48] J. Goebel and T. Holz, "Rishi: Identify bot contaminated hosts by irc nickname evaluation," in *Proceedings of the first conference on First*

Workshop on Hot Topics in Understanding Botnets, ser. HotBots'07. Berkeley, CA, USA: USENIX Association, 2007, pp. 8–8.

[49] G. Gu, J. Zhang, and W. Lee, "BotSniffer: Detecting botnet command and control channels in network traffic," in *Proceedings of the 15th Annual Network and Distributed System Security Symposium (NDSS '08)*, February 2008, pp. 1–1.

[50] A. Ramachandran, N. Feamster, and D. Dagon, "Revealing botnet membership using dnsbl counter-intelligence," in *Proceedings of the 2nd conference on Steps to Reducing Unwanted Traffic on the Internet - Volume 2*, ser. SRUTI'06. Berkeley, CA, USA: USENIX Association, 2006, pp. 8–8.

[51] R. Villamarin-Salomon and J. Brustoloni, "Identifying botnets using anomaly detection techniques applied to DNS traffic," in *Proceedings of 5th IEEE Consumer Communications and Networking Conference (CCNC 2008)*, 2008, pp. 476–481.

[52] R. Villamarín-Salomón and J. C. Brustoloni, "Bayesian bot detection based on dns traffic similarity," in *Proceedings of the 2009 ACM symposium on Applied Computing*, ser. SAC '09. New York, NY, USA: ACM, 2009, pp. 2035–2041.

[53] X. Yu, X. Dong, G. Yu, Y. Qin, D. Yue, and Y. Zhao, "Online botnet detection based on incremental discrete fourier transform," *JNW*, vol. 5, no. 5, pp. 568–576, 2010.

[54] T. M. Mitchell, *Machine Learning*, 1st ed. New York, NY, USA: McGraw-Hill, Inc., 1997.

[55] S. Kotsiantis, I. Zaharakis, and P. Pintelas, "Supervised machine learning: A review of classification techniques," *Frontiers in Artificial Intelligence and Applications*, vol. 160, p. 3, 2007.

[56] A. K. Jain, M. N. Murty, and P. J. Flynn, "Data clustering: A review," *ACM Comput. Surv.*, vol. 31, no. 3, pp. 264–323, Sep. 1999.

[57] Suricata, IDS, "open-source ids/ips/nsm engine," 2015.

[58] N. Provos and T. Holz, *Virtual honeypots: From botnet tracking to intrusion detection*, 2nd ed. Addison-Wesley Professional, 2009.

[59] E. Stinson and J. C. Mitchell, "Towards systematic evaluation of the evadability of bot/botnet detection methods," in *Proceedings of the 2nd conference on USENIX Workshop on offensive technologies*, ser. WOOT'08. Berkeley, CA, USA: USENIX Association, 2008, pp. 5:1–5:9.

[60] T. T. Nguyen and G. Armitage, "A survey of techniques for internet traffic classification using machine learning," *IEEE Communications Surveys & Tutorials*, vol. 10, no. 4, pp. 56–76, 2008.

[61] M. Kührer, C. Rossow, and T. Holz, "Paint it black: Evaluating the effectiveness of malware blacklists," in *Research in Attacks, Intrusions and Defenses*. Springer, 2014, pp. 1–21.

[62] C. J. Dietrich and C. Rossow, "Empirical research of ip blacklists," in *ISSE 2008 Securing Electronic Business Processes*. Springer, 2009, pp. 163–171.

[63] S. Sinha, M. Bailey, and F. Jahanian, "Shades of grey: On the effectiveness of reputation-based âĂİJblacklistsâĂİ," in *3rd IEEE International Conference on Malicious and Unwanted Software, MALWARE 2008*. 2008, pp. 57–64.

[64] S. Sheng, B. Wardman, G. Warner, L. Cranor, J. Hong, and C. Zhang, "An empirical analysis of phishing blacklists," in *Sixth Conference on Email and Anti-Spam (CEAS)*, 2009.

[65] N. Kheir, F. Tran, P. Caron, and N. Deschamps, "Mentor: Positive dns reputation to skim-off benign domains in botnet c&c blacklists," in *ICT Systems Security and Privacy Protection*. Springer, 2014, pp. 1–14.

Biographies

M. Stevanovic received the M.Sc. in Electrical Engineering in 2011, from the Faculty of Electrical Engineering, Belgrade University, specializing in system engineering. He is currently a Ph.D. Student in the Wireless Communication Section, Department of Electronic Systems, Aalborg University. His research interests include network security, traffic anomaly detection and malware detection based on network traffic analysis.

J. M. Pedersen received the M.Sc. in Mathematics and Computer Science in 2002, and the Ph.D. in Electrical Engineering in 2005 from Aalborg University, Denmark. He is currently Associate Professor at the Wireless Communication Section, Department of Electronic Systems, Aalborg University. His research interests include network planning, traffic monitoring, and network security. He is author/co-author of more than 70 publications in international conferences and journals, and has participated in Danish, Nordic and European funded research projects. He is also board member of a number of companies within technology and innovation.

Practical Attacks on Security and Privacy Through a Low-Cost Android Device

Greig Paul and James Irvine

University of Strathclyde Department of Electronic & Electrical Engineering
Glasgow, United Kingdom
Email: {greig.paul; j.m.irvine}@strath.ac.uk

Received 3 November 2015; Accepted 1 December 2015;
Publication 22 January 2016

Abstract

As adoption of smartphones and tablets increases, and budget device offerings become increasingly affordable, the vision of bringing universal connectivity to the developing world is becoming more and more viable. Nonetheless, it is important to consider the diverse use-cases for smartphones and tablets today, particularly where a user may only have access to a single connected device. In many regions, banking and other important services can be accessed from mobile connected devices, expanding the reach of these services.

This paper highlights the practical risks of one such low-cost computing device, highlighting the ease with which a very recent (manufactured September 2015) Android-based internet tablet, designed for the developing world, can be completely compromised by an attacker. The weaknesses identified allow an attacker to gain full root access and persistent malicious code execution capabilities. We consider the implications of these attacks, and the ease with which these attacks may be carried out, and highlight the difficulty in effectively mitigating these weaknesses as a user, even on a recently manufactured device.

Keywords: Security, Privacy, Android, Exploit, Physical Access.

Journal of Cyber Security, Vol. 4, 121–140.
doi: 10.13052/jcsm2245-1439.422
ⓒ 2016 *River Publishers. All rights reserved.*

1 Introduction

The smartphone and tablet markets in developing regions are predicted to see significant growth in the coming years. Time magazine reported on the current low market penetration of smartphones in India, and the likelihood for rapid growth in sales of lower price-point handsets [1]. Cisco's Visual Networking Index predicts that by 2019, mobile data traffic usage in India will grow 13-fold from 2014, and that 51% of devices will be smart devices (i.e. smartphones and tablets) rather than feature-phones [2].

In Africa, there has been a widely documented rise in the use of mobile phones by individuals, as both a means of communication, as well as a means of access to services from governments and banks [3] [4, pp. 80–90]. The rise of mobile banking in Africa is also especially significant, as it has reached significant market penetration and day-to-day usage, with mobile payments in Africa exceeding those from both Europe and North America [5, 6].

There is therefore a motivation for technology-based attacks against users in these regions, leading towards crimes such as theft. This paper aims to explore some of the risks posed by consumer electronic devices targeted specifically towards developing markets. We highlight the risks posed to users as a result of some of the security weaknesses in these devices, and consider the challenges of providing secure computing environments in these situations.

Conventionally in the study of security, attacks involving compromise of a local device are generally neglected, on account of the adage that physical access to a device grants the attacker control over the device. Despite this, the threat model for consumer hardware is rapidly evolving, with Secure Boot in UEFI on desktop and laptop computers able to protect against malicious modifications to the computer's operating system during offline attack [7]. For this reason, we consider physical attacks from non-invasive attacks (i.e. not requiring physical disassembly of the product) to be highly relevant and significant to users of these devices, and shall illustrate the significance of these attacks in Section 4.

2 Overview of Android Security Model

The Android operating system is derived from an open source project, referred to as the Android Open Source Project (AOSP) [8]. Android is a customised userland, built on top of the Linux kernel, providing APIs for sandboxed applications to run, and access hardware in a standardised manner, irrespective of the manufacturer of the device. This means that many devices run the Android operating system, from different manufacturers, and application software may run on any of these, using the Android platform's hardware abstraction to simplify the development of software.

Android devices, in contrast to more general purpose Linux-based computers, are designed around a security model whereby each user application runs within a separate Unix user account (UID). By controlling access to protected resources through a combination of a permissions layer provided by the Android APIs, and regular filesystem permissions, applications are sandboxed from each other, and are not normally able to access data from other applications.

This isolation offers a different security model from the desktop. The Android model assumes that an application may safely store data privately within its own dedicated area, protected by the filesystem access control permissions. This prevents other applications, or indeed the user themselves, from having direct access to these files.

Android devices, in common with most Linux-based embedded systems, feature a number of separate partitions, in order to separately hold the kernel image, ramdisk, and main operating system. There is additionally a partition to hold the user's data, which is able to be erased in the process of carrying out a factory reset. Ordinarily, no partition other than the user data partition should be mounted read-write — the operating system partition is, by default, mounted read-only, and other partitions are not generally mounted, since they do not contain regular filesystems.

In line with other Unix-based systems, root access is the highest level of privilege available to code running under the kernel. Code executed as the root user may load modules into the kernel, altering its behaviour, and generally access any resource provided by the kernel, with full access permissions.

2.1 Main Requirements

The Android compatibility definition document, as published by Google [9], is intended as a guide as to good practice when making a device which runs the Android operating system. In order to receive approval for the shipping of Google services on a device (which we note the device investigated in this paper did not include), it is necessary for these guidelines to be followed.

On Android devices, the root user account is not meant for normal use, and there is not meant to be a means for the end user (or applications) to access it [9]. This is because the root user is inherently able to access the entire filesystem, thus bypassing the access controls of the Android security model, and this would expose protected application data to any code running as root.

Additionally, the Android compatibility definition now requires the use of SELinux in enforcing mode [9]. SELinux is a mandatory access control framework, designed to enforce system-wide security policies across a system. It is capable of constraining applications which run as the root user, and therefore offers some protection against the use of kernel root exploits and other privilege escalation attacks, since such an exploit will not bypass the constraints of the SELinux policies [10]. This naturally requires well-designed and implemented policies, since SELinux merely enforces those policies.

3 Overview of Platform Hardware

While the Android operating system is shipped on a wide variety of devices, at a wide variety of price points, we focus exclusively on low-cost, consumer hardware. The tablet in question we investigated was manufactured in September 2015, according to the information label on the box, shipped with Android 4.4 (KitKat), and retailed for 3499 Rupees in India [11] (approximately 50 USD). It was distributed to attendees of meeting 35 of the Wireless World Research Forum, as a conference gift holding the proceedings.

Android devices feature a bootloader, which initialises the hardware, loads the kernel from storage into memory, and executes the kernel, having set the kernel commandline parameters to specify where to locate the remainder of the operating system.

In addition to loading the kernel, the bootloader is also used to control boot modes, based upon the boot control block, which is implemented to allow a device to boot into a minimal recovery environment. This environment normally does not require the Android userland to be present to run, and ships with its own kernel image and ramdisk, from which the main operating system and kernel can be updated. This environment also carries out factory reset operations, erasing the user data partition. The bootloader also typically offers a means of reprogramming a device's internal memory, to allow for the recovery image to be upgraded, and to allow the device to be programmed as part of the manufacturing process.

4 Vulnerability Identification

In this section, the process through which vulnerabilities were identified on the device is discussed, as well as the capabilities of each exploit. These vulnerabilities are reported in the order they were originally identified during our research. Finally, following our exploration of these identified issues, we carried out a security scan of the underlying Android operating system (since our identified vulnerabilities mostly lie below the Android operating system), and highlight the risks identified there.

4.1 Root ADB Access by Default

Initial exploration of the device indicated that ADB (Android Debugging Bridge) was enabled by default, which is a property of engineering and userdebug builds (referred to as "eng" or "userdebug" builds) of Android. These builds feature reduced security, compared with release firmware (referred to as "user" builds). This was confirmed by carrying out a factory reset of the device, and verifying that ADB remained enabled by default. The build fingerprint was also checked using getprop ro.build.type, indicating the build was a "userdebug" build.

One of the relevant security features disabled in non-release builds is ADB host verification, which requires the user accept a public key presented by the computer opening a connection with the ADB daemon on the phone. This meant that it was possible for an ADB session to be established without either a prompt being shown to the user, or confirmation being given by the user.

The ADB connection available over USB offered a standard Unix shell on the device, from which commands may be executed by any device connected to the USB port. One of the binaries available on the device was the su binary, designed to escalate the current user to root. On this device, it was possible to carry out an escalation from the ADB shell user, to the root user, without any prompt of input. This escalation to root access was confirmed using the Unix id command, indicating the shell was running as UID 0, that of the root user.

It was not possible for user applications to carry out a root escalation using this approach directly, since root access was only granted to the shell user. Nonetheless, this poses a risk for users charging or otherwise plugging their device into an untrusted charger, or where others may have even momentary physical access to the device. In many developing countries, where access to grid-supplied electricity is not practical, users charge devices in shops or public charging stations [12], putting them at risk of a rogue charger connecting over ADB and gaining root access to the device.

4.2 SELinux Bypass

Although root access was obtained from the ADB shell, this access was still potentially subject to SELinux policy constraints. While no denials were encountered in the process of carrying out this work, it was found to be possible to easily disable SELinux. SELinux was firstly ascertained to be enabled through the use of the `getenforce` command, which indicated the policies were in "Enforcing" mode (rather than merely in permissive fault logging mode). By running the command `setenforce 0` from the rooted shell, it was possible to disable the SELinux access controls, as verified by the "Permissive" response from the `getenforce` command.

4.3 Bootloader Root Shell

Access to the device bootloader was gained by holding down both the volume down, and power buttons, to turn on the device. In this mode, the device presented a menu of options, selectable using the volume keys and power button. In bootloader mode, an ADB device was again presented over the USB interface, once again without ADB authentication. Upon opening the shell, it was possible to escalate to root access using the `su` command. The presence of this vulnerability in the bootloader means that even in the event of the regular firmware being patched or upgraded, it would also be necessary to make significant modifications to the lower level boot stack, which may or may not be practical to carry out, given the risks of carrying out an upgrade of device bootloaders in-the-field.

While accessing this shell, it was observed that the user data and operating system filesystems were both mounted in read-write mode. This meant that it was possible to easily make persistent modifications to the operating system image, or access user data, directly from the bootloader shell.

We also observed that the bootloader installed on the device to the partition `mmcblk0boot0` did not feature a cryptographic signature at its footer, indicating it is likely that this bootloader is unsigned, and therefore potentially vulnerable to tampering or modification by a suitably determined adversary.

4.4 Recovery ZIP Signing Keys

The recovery environment, used to install operating system updates, has the capability for a ZIP file containing new firmware to be loaded into memory and installed. This ZIP file should be signed with a private key corresponding to a certificate stored on the device itself, in order to verify that the firmware being installed has not been modified or corrupted in the process of reaching the device.

This signature check relies upon the confidentiality of the firmware zip signing keys — if a third party is able to generate signed firmware images, they may replace any component of the device operating system, including installed applications or even the device kernel, simply through creation of a custom firmware ZIP file. In the case of this device, the recovery image accepted standard ZIP signing keys, which are publicly available within the AOSP source code repositories [13]. As a proof of concept, a ZIP was created to display a message to the screen, and add a new file to the device filesystem. It was then signed using the AOSP testkey, and successfully installed onto the device, using the "Install from ADB" feature of the recovery environment.

The file `input.zip` was signed using the command `java -jar signapk.jar -w testkey.x509.pem testkey.pk8 input.zip output -signed.zip`, and installed using the command `adb sideload output-signed.zip`

The execution of the ZIP file was confirmed through the output of the command to display a message to the screen, and the new file added to the filesystem being observed following a reboot. This illustrated it was possible to make arbitrary modifications to the device operating system, such as adding new files, or modifying existing files, which would not be reverted following a device factory reset.

4.5 Application Package Signing Keys

Android applications (APKs) are signed in a similar manner to ZIP firmware upgrades. To prevent application replacement attacks, where a user is encouraged or coerced into installing a false update to an application, the Android platform will not allow an update to an application to be installed if its signing key does not match the existing signing key for the application. This ensures that the party signing the APK holds the same key as originally used by the developer. Likewise, applications are protected against downgrading, by ensuring that the version code has been incremented since the previous update, which could be used to install an old version of an application with vulnerabilities, for exploitation.

The signing keys used for the platform applications (which have privileged access to system APIs) were found to, again, be the default AOSP signing keys, which are publicly available. It was therefore possible to replace core system applications on the device, including for example, the dialer, settings interface, keyboard, and overall firmware user interface (referred to as SystemUI).

A modified version of these applications could then be used to upgrade an existing version of the application, without the user being made aware of any extra risks. This is of particular significance since platform level applications have full access to the entire device and permissions. Having a publicly available platform key significantly violates the Android security model, which assumes the platform key is not available to attackers [14].

4.6 Android OS Security Status

To conclude our analysis of the device, we carried out an analysis of the device's resistance to a variety of standard, well-known attacks and exploits against the Android operating system, using the Bluebox's "Trustable" security scanner [15]. The results of this highlighted that the device was protected against only 3 of the 12 vulnerabilities scanned for. This scanner was selected, as it is capable of detecting all of the recent high-profile security vulnerabilities of the Android operating system, including "StageFright", the multiple variants of the "MasterKey" attack, and a number of kernel root exploits, including the futex attack. Figure 1 highlights the results of this scan. We also note that the device was vulnerable to the CVE-2015-3636 local privilege escalation attack via kernel ping sockets. An open source implementation of this exploit is available [16], and was used to verify that the device was vulnerable. Any application capable of executing a binary on this device (or indeed a user with access to a shell) was able to gain local root access, as shown in Figure 2, where a shell running as the root user was obtained.

Of these vulnerabilities, FakeID, Futex, ObjectInputStream and Pending-Intent were reported in 2014, yet remained un-patched in this device, with a manufacture date of September 2015. This was due, in part, to some of these fixes being withheld until the release of future major versions of Android, rather than immediate security patches being released and backported to older software versions. While Google has begun to issue security backport patches and notifications [17], this is a very recent change, and requires the vendor to apply these patches. In the case of this device, the presence of serious weaknesses like the futex root exploit, suggest this is not the case, and that patches are not being applied prior to the launch of devices.

Indeed, by checking the build date of the software on the device from the command `getprop ro.build.date`, the software was found to have

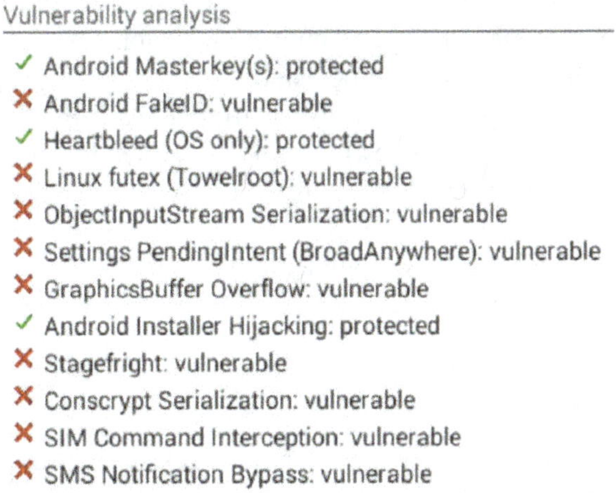

Vulnerability analysis

✓ Android Masterkey(s): protected
✗ Android FakeID: vulnerable
✓ Heartbleed (OS only): protected
✗ Linux futex (Towelroot): vulnerable
✗ ObjectInputStream Serialization: vulnerable
✗ Settings PendingIntent (BroadAnywhere): vulnerable
✗ GraphicsBuffer Overflow: vulnerable
✓ Android Installer Hijacking: protected
✗ Stagefright: vulnerable
✗ Conscrypt Serialization: vulnerable
✗ SIM Command Interception: vulnerable
✗ SMS Notification Bypass: vulnerable

Figure 1   Result of Bluebox security scan for Android vulnerabilties.

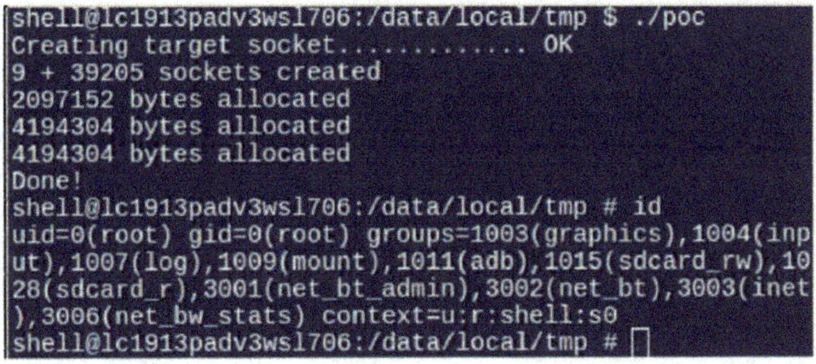

```
shell@lc1913padv3wsl706:/data/local/tmp $ ./poc
Creating target socket............. OK
9 + 39205 sockets created
2097152 bytes allocated
4194304 bytes allocated
4194304 bytes allocated
Done!
shell@lc1913padv3wsl706:/data/local/tmp # id
uid=0(root) gid=0(root) groups=1003(graphics),1004(inp
ut),1007(log),1009(mount),1011(adb),1015(sdcard_rw),10
28(sdcard_r),3001(net_bt_admin),3002(net_bt),3003(inet
),3006(net_bw_stats) context=u:r:shell:s0
shell@lc1913padv3wsl706:/data/local/tmp #
```

Figure 2   Successful execution of local root exploit CVE-2015-3036.

been built in April 2015, several months before the product's release date. The software therefore appeared to not have been rebuilt by the manufacturer in the 5 months prior to launch, thus explaining the lack of many security patches. Since the device in question had no over-the-air update capabilities, we suggest a device shipping with 5 months' of disclosed vulnerabilities present puts users at risk from the moment the device is removed from its box.

4.7 Summary of Attack Vectors

A number of vulnerabilities have been identified on this device. All of these were in the default configuration, in an out-of-box setup as experienced by users. They are listed below in summary form.

- Privilege escalation to root possible from ADB shell on device
- ADB shell accessible without authentication, enabled by default
- SELinux can be disabled via a root shell
- ADB shell allowing privilege escalation to root available in bootloader
- Recovery image uses well-known signing keys intended only for testing
- Platform applications are signed using well-known signing keys intended only for testing
- Multiple previously disclosed vulnerabilities, including root exploits, unpatched on released device.

5 Potential Attacks

This section shall consider some of the potential attacks which could be carried out against users, as a result of the attack vectors described previously. These attacks should be considered in the context of a user in the developing world, who may be using this device for a variety of tasks, including banking or accessing government services, as discussed previously.

Previous works have examined user attitudes towards the perceived sensitivity and value of personal data on smart-phones, and what concerned them most about potential actions of software. A survey of these actions highlighted that, asides from permanently disabling or breaking a handset, the top ten concerns from users related to actions which would cost the user money (such as making premium rate calls or sending premium rate SMS messages), or were destructive (such as deleting user data like contacts) [18]. Other concerns raised included the public sharing of data which users felt was confidential, such as their text messages, emails, or photographs. Users were also concerned about the sharing of their data with advertisers, or the abuse of their contacts for spamming, and the risk of theft of passwords or other credentials such as credit card details.

In a survey of perceived value and sensitivity of their data, the sensitivity of location data and passwords was highlighted, as well as that of other types of data, such as photos and videos, or of messages [19]. It is therefore clear that users are concerned both about the theft of their data, and of the risk of the loss of such data, and the consequence of its loss.

5.1 Lockscreen Bypass

The first attack we identified allows an unauthenticated attacker to bypass the device's lockscreen, if the user made use of a PIN, password or pattern lock for security. Since the lockscreen on Android is designed to fail insecure, in the event of corrupted (or missing) settings, simply removing the pattern or PIN data is sufficient to completely bypass the lockscreen. By removing the file /data/system/password.key, or /data/system/gesture.key, for PIN/password or pattern locks respectively, the lockscreen security was completely removed. Alternatively, the cryptographic hash of the password may be obtained from this file, and brute-forced, as described in [20, p. 268–275], in order to establish the plaintext password, PIN, or pattern lock, as set by the user.

This attack is possible, since the device exposes an unauthenticated ADB shell by default, with privilege escalation to root available through use of the su command. In the event that ADB is disabled on the device, it may also be carried out directly from the root ADB shell available through the bootloader. Finally, a firmware update ZIP file could be crafted (and signed) to remove this file via recovery, thus removing lockscreen security, and providing the user with full access to the device as though a password had been entered.

The ability to bypass the lockscreen poses a clear privacy risk to users, as it serves as an authentication bypass to carry out operations such as accessing user data, sending messages, and similar. It also allows access to messages and potentially sensitive files, including (for example), one-time passwords sent over SMS to the device. The ability to bypass the lockscreen therefore aids an attacker in carrying out interactive exploration of the device.

5.2 Theft of User Data

With ADB access available to the device by default, the adb pull command may be used to extract files from the shared storage area of the device, or the SD card. These areas of storage are not protected by per-application filesystem permissions, and are designed for the storage of data which a user may wish to transfer to a computer. The adb pull command does not require root access to succeed, and may be carried out either from within the regular operating system, or from the bootloader. In addition, due to the lack of authentication on ADB connection attempts, it is also possible to use ADB from the lockscreen to access data.

This attack makes it possible for a malicious attacker to extract all of a user's photographs, or documents, from the device. By expanding this attack to utilise root permissions (through privilege escalation via the su command), it is possible to further extend it to result in the theft of private per-application user data. This may include passwords for user accounts, as well as user messages, photographs, cryptographic keys, and other sensitive data which should only be accessible to a single application, such as tokens. If this were carried out against a banking app, there is potential for sensitive user data to be obtained, depending upon the design of the application in question.

It was also possible to gain a full image of the device's data partition, using root access to recursively select all files found on the user data partition, and store them in a single compressed archive, which could easily be extracted from the device over ADB.

5.3 Full Access to Device Partitions

With full read-write access to the device, including operating system partitions, it was possible to make modifications to the installed operating system. This was achieved from within the operating system itself, by re-mounting the system partition as root using the command mount -o remount, rw /system. It was also possible to carry this out from the bootloader shell, as well as through the installation of a custom ZIP from the recovery environment. These changes are persistent through a factory reset, increasing the severity of this attack, since rectifying the modifications will be beyond the abilities of most users, if they were able to detect the modifications in the first place.

5.4 Installation of a Keylogger

A malicious attacker may wish to capture user credentials, in order to gain access to user accounts, or financial credentials such as card details. Alternatively, they may simply wish to know what a user is writing on their device, especially if it is used to carry out sensitive tasks. It was possible to install a keylogger through a variety of methods on this device, without the user being aware. Firstly, from a rooted shell, it was possible to replace the default keyboard application on the system partition (this change will persist between installs). It was also possible to remotely socially engineer a user to install an updated version of the keyboard application, using the publicly known signing keys to create an apparent update to the keyboard. This update could be distributed on the internet, or indeed installed through the USB interface of the device while charging. An updated version of an application could be installed over ADB using the command adb install -r filename.apk, or directly from the device filesystem by downloading it to the shared storage, and selecting the application from the file manager.

5.5 Deletion of Data

With root access to the device, it was possible to gain full access to the filesystems, and to erase all user data. By carrying out a backup prior to this process, and potentially encrypting the backup, it would be possible for a malicious party to ransom a user's data, requiring them to make a payment to gain access to it again. As one of the main concerns of users was the deletion of their data, this may cause significant inconvenience to users [18]. The ability to scale this attack to many devices, through the use of communal charging areas [12], would also make the ability to spread malicious software like this particularly harmful.

5.6 SMS Interception

With root access to the device, it is possible for the device operating system to be modified, such that SMS messages may be intercepted or redirected. This could take place without the user being notified or aware, and could be used to relay seemingly secure one-time passwords, which are commonly sent via SMS. This modification would be made in the messaging app or system frameworks, and could be carried out either using root access from a plugged-in device, or by socially engineering a user to install a rogue "updated" version of the messaging app, using the publicly known signing keys to sign the application package.

5.7 Premium Rate Abuse

One of the main concerns identified in [18] was the risk that rogue software could run up a bill by making premium rate calls or text messages. With root-level access to Android devices, it is possible to modify the dialer to make calls when the device is not in use, or to modify the messaging app to silently (without a record kept in the sent items) subscribe the user to premium rate SMS services, and send premium rate messages. While the Android operating system contains software to warn the user before premium rate SMS messages are sent, these warnings are easily removed or bypassed on a rooted device, where the operating system may be modified without user intervention or knowledge. A financially motivated attacker with the ability to carry out this attack on many devices, such as by creating a fake mobile phone charging station, could potentially compromise a large number of devices, and make significant quantities of money by making use of premium rate network services on behalf of unwitting users. To hinder discovery of such an attack, the modification may be configured to only operate when the device is charging, such that no abnormal power use would be noticed by a user.

5.8 Random Number Generation Compromise

Another attack which we found to be possible was the effective "short-circuiting" of the kernel's random number generation. This was made possible through the root access exposed by the device. By renaming /dev/urandom, and creating a new symbolic link towards /dev/zero using the command ln -s /dev/zero /dev/urandom, all random number generation using the kernel APIs was compromised, after the depletion of the existing random pool by running cat /dev/random. A long series of random numbers was generated, and found to return all-zeroes, as expected. The ability for an attacker to do this allows for the compromise of security of encrypted communications, since randomly generated keys would be entirely predictable. Additionally, numbers expected to be random (such as nonces or initialisation vectors), may be re-used and predictable, potentially compromising the security of the protocol. In particular, for elliptic-curve based signature algorithms such as ECDSA, re-use of a nonce causes a catastrophic failure resulting in the ability for the signing key to be determined from 2 signatures using the same nonce [21, p. 68–72]. There is therefore potential for significant harm, if malicious software, or indeed a malicious attacker, were to abuse root access in order to render ineffective random number generation on the device. This may also expose a user to further attacks, since private components generated for Diffie-Hellman key exchange (as seen in TLS), would be predictable and repeated, potentially resulting in the ability for interception of HTTPS-based traffic. On a device where sensitive tasks, such as mobile banking, were being carried out, it would then be possible for a remote attacker with ability to observe network traffic (such as over an insecure Wi-Fi network) to break an HTTPS connection through knowledge of the compromised output of the random number generator.

5.9 Physical Damage to Device

Finally, with root access, it would be possible for a malicious attacker to abuse root access to render the device inoperable, and useless, effectively depriving the user from their computing device. By using the Unix dd command, it is possible to carry out raw read and write operations to block devices on the device. By overwriting the kernel image and recovery image, the device would be beyond the repair capabilities of most users, since there would be no standard means through which to restore the firmware. As we noticed the device bootloader was unsigned, it is likely that an attacker could overwrite this with an empty partition, to permanently prevent the device from working. While there may be no clear or strong motive to cause physical damage to user devices, the ability to carry out this attack is a concern — potentially an unscrupulous retailer could try to drive more sales of new devices by having an attacker abuse this access to cause damage to devices. For a user reliant upon their mobile device as their main means of accessing electronic services, this would be potentially disruptive, and also cause them to lose data, reducing their confidence and trust in the technology.

6 Potential Mitigations

While the fundamental vulnerabilities identified here cannot necessarily be resolved easily by end-users, it is possible to identify some potential countermeasures to take. Firstly, users should disable ADB on the device, from the "Developer options" menu. This setting is ordinarily disabled by default, but was enabled on the device due to an engineering build of firmware being shipped on the finished device. Secondly, users should consider making use of a USB cable with the data lines shorted together, to prevent the transfer of data across the USB port when charging. Since the bootloader may be attacked separately from the operating system, users should ensure the device is always charged using this cable, and that the device is kept in sight at all times while charging (to prevent the cable being swapped). While there is no easy way to prevent abuse of the publicly-known signing keys used for the recovery image and platform applications, vigilance against installing any third party software, and avoiding the installation of any software would help to alleviate this. This would be a considerable trade-off for users, however, to forego the installation of software in order to avoid such attacks.

A technically confident user may attempt to remove the su binary from the device by remounting the system as read-write, as discussed previously, and using the command `rm /system/xbin/su`, although this will not protect the device from other attacks such as the abuse of known recovery ZIP and platform application signing keys, or indeed the use of the CVE-2015-3036 root exploit, as demonstrated in Section 4.6. This procedure is also somewhat risky in that removing the wrong file may result in the device being unbootable.

On future devices, it would be beneficial for over-the-air firmware updates to be possible as well — on this device, there was no facility inbuilt for network-based updates that we could identify. Instead, there was a menu option inviting the user to place a file named `update.zip` on the SD card, and select a menu option to install it. In addition to the problem of the signing key being known for these update files, making it easy for a malicious party to distribute fake updates containing malicious software, the manufacturer is unable to issue prompt and regular security updates directly to devices, to address issues identified.

7 Conclusions

In this paper, we have highlighted some major security weaknesses in a recent, low-cost Android device, intended for developing markets. We identified that insufficient measures were in place to protect user data. We showed how a root shell could be gained on the device, out of the box, through both the device firmware and the bootloader. We also demonstrated the device shipped with a kernel root vulnerability, and that this is exploitable by any locally running software (such as an app). We also showed how persistent modifications to the firmware could be made, which would persist through factory resets, allowing for highly pervasive malicious software to be installed and target user data. Furthermore, we highlighted the risks of the device using the default firmware signing keys, and application signing keys, the private components of which are publicly available. We demonstrated that SELinux mandatory access control could easily be disabled by the root user.

The implications of vulnerabilities such as this are particularly significant for users of devices in the developing world, often the recipients and buyers of such low-cost devices as their main computing device. Merely plugging this device into a public charging station would be sufficient for a malicious party to gain full control over the device, extract all of a user's personal data, and install pervasive malicious software onto the device. This software could act as a keylogger, recording passwords and financial information, or could serve to silently gather sensitive data (such as two-factor authentication SMS messages) and forward them to the attacker. Malicious software could also erase all of a user's data, and demand a ransom for its return, or even simply destroy the device. We demonstrated these changes will persist through a factory reset, and that they are not visible to the end user. As the changes persist, it is not possible for a user to remove such malicious changes without advanced technical knowledge, and a known-clean firmware image to replace their device's software with. While we have presented some mitigations against these attacks, these require the user to be highly vigilant, and would not allow them to make use of many of the functions of the device (such as installing applications), in order to defend against some of these attacks.

With the rise in adoption of smartphones and tablets in developing markets, the security and privacy of their users should be considered a priority when developing their software. Guidelines from the Android Compatibility Definition documents should be followed to avoid known security weaknesses. Finally, low level device firmware should be audited to ensure that bootloaders and other interfaces do not expose low-level root access to a device by default, which would undermine the security model of an otherwise-secure operating system.

Acknowledgment

This work was funded by EPSRC Doctoral Training Grant EP/K503174/1.

References

[1] B. Bajarin. (December 2014) Why India will be the world's second biggest smartphone market. [Online]. Available: http://time.com/3611863/india-smartphones/

[2] Cisco. (May 2015) VNI mobile forecast highlights, 2014–2019. [Online]. Available: http://www.cisco.com/assets/sol/sp/vni/forecast_highlights_mobile/index.html

[3] S. Etzo and G. Collender, "The mobile phone revolution in Africa: Rhetoric or realty?" *African affairs,* 2010.

[4] K. E. Skouby and W. Idongesit, *The African Mobile Story.* River Publishers, 2014.

[5] D. Porteous, "The enabling environment for mobile banking in Africa," 2006.

[6] B. Warner. (March 2013) What Africa can teach us about the future of banking. [Online]. Available: http://www.bloomberg.com/bw/articles/2013-03-06/what-africa-can-teach-us-about-the-future-of-banking

[7] G. Paul and J. Irvine, "Take control of your PC with UEFI secure boot," *Linux J.,* vol. 2015, no. 257, Sep. 2015.

[8] Google. The android source code. [Online]. Available: http://source.android.com/source/

[9] Google. (October 2015) Android 6.0 compatibility definition. [Online].
     Available: http://source.android.com/compatibility/index.html

[10] S. Smalley and T. M. R2X, "The case for SE Android," *Linux Security Summit*, 2011.

[11] (October 2015) Datawind Ubislate 27CZ. [Online]. Available: http://www.pricedealsindia.com/mobiles/Datawind-Ubislate-27CZ-price-in-india-dpi4016.php#gotostore

[12] D. Wogan. (November 2013) Charging a mobile phone in rural Africa is insanely expensive. [Online]. Available: http://blogs.scientificamerican.com/plugged-in/charging-a-mobile-phone-in-rural-africa-is-insanely-expensive/

[13] Google. (October 2008) AOSP platform signing keys. [Online]. Available: https://android.googlesource.com/platform/build/+/master/target/product/security/

[14] D. Hackborn. (May 2011) Restrict system packages to protected storage, android code review. [Online]. Available: https://android-review.googlesource.com/#/c/22694/

[15] J. Forristal. (October 2014) Measuring mobile security & trust: Introducing trustable by bluebox. [Online]. Available: https://bluebox.com/measuring-mobile-security-trust-introducing-trustable-by-bluebox/

[16] CVE-2015-3636. Commit used: 9868289bdb53c. [Online]. Available: https://github.com/fi01/CVE-2015-3636

[17] Google. (August 2015) Android security updates. [Online]. Available: https://groups.google.com/forum/#!forum/android-security-updates

[18] A. P. Felt, S. Egelman, and D. Wagner, "I've got 99 problems, but vibration ain't one: a survey of smartphone users' concerns," in *Proceedings of the second ACM workshop on Security and privacy in smartphones and mobile devices*. ACM, 2012, pp. 33–44.

[19] I. Muslukhov, Y. Boshmaf, C. Kuo, J. Lester, and K. Beznosov, "Understanding users' requirements for data protection in smartphones," in *Data Engineering Workshops (ICDEW), 2012 IEEE 28th International Conference on*. IEEE, 2012, pp. 228–235.

[20] N. Elenkov, Android Security Internals: *An In-Depth Guide to Androids Security Architecture*. San Francisco: No Starch Press, 2015.

[21] A. Kak, "Elliptic curve cryptography and digital rights management," *Lecture Notes on Computer and Network Security*, 2015. [Online]. Available: https://engineering.purdue.edu/kak/compsec/NewLectures/Lecture14.pdf

Biographies

G. Paul received the B.Eng. (Hons.) degree in Electronic & Electrical Engineering from the University of Strathclyde, Glasgow, UK, in 2013. He is currently pursuing the Ph.D. degree in the Mobile Communications Group at the University of Strathclyde. He is a Graduate Student Member of the IEEE, and The Institution of Engineering and Technology, and the Chair of the University of Strathclyde IEEE Student Branch. His research interests include secure data storage and retrieval, practical considerations in the design of secure systems, and the design of privacy-preserving service architectures. Greig is the recipient of an EPSRC Doctoral Training Grant.

J. Irvine received the B.Eng. (Hons.) degree in Electronic and Electrical Engineering and the Ph.D. degree in coding theory from the University of Strathclyde, Glasgow, U.K., in 1989 and 1994, respectively. He is currently a Reader with the Department of Electronic and Electrical Engineering, University of Strathclyde, Glasgow, U.K., where he also leads the Mobile

Communications Group. He is a coauthor of seven patents and the books Digital Mobile Communications and the TETRA System (Wiley, 1999) and Data Communications and Networks: An Engineering Approach (Wiley, 2006). His research interests include mobile communication and security, particularly resource allocation and coding theory. Dr. Irvine is an elected member of the Board of Governors of the IEEE Vehicular Technology Society, a member of the IET, a Fellow of the Higher Education Academy, and is a Chartered Engineer.

Comparative Investigation of ARP Poisoning Mitigation Techniques Using Standard Testbed for Wireless Networks

Goldendeep Kaur and Jyoteesh Malhotra

Computer Science and Engineering Department, Guru Nanak Dev University,
Regional Campus, Jalandhar
E-mail: goldenchugh@gmail.com; jyoteesh@gmail.com

Received 4 September 2015; Accepted 25 September 2015;
Publication 22 January 2016

Abstract

Due to the increasing demand of wireless networks, there is an increasing necessity for security as well. This is because unlike wired networks, wireless networks can be easily hacked form outside the building if proper security measures are not in place as wireless networks make use of radio waves and radio waves can leak outside of building at distances up to 300 feet or more. So everything we do on our network can be monitored by anyone who has wireless capabilities. This unauthorized access can be used as an essence by the hacker to launch various kinds of attacks like man-in-the-middle attacks, denial of service attacks, IP spoofing etc. As a result in addition to the firewalls, password protection techniques, virus detectors etc, additional levels of security is needed to secure the wireless networks. This paper focuses on comparing various techniques that are used to protect the users from these attacks by providing practical observations based on the network parameters time and scalability and also highlighted the best method in the end to combat the attacks at a superior level.

Keywords: ARP Cache, Snort, Wireshark, SSLstrip, Ettercap.

Journal of Cyber Security, Vol. 4, 141–152.
doi: 10.13052/jcsm2245-1439.423
© 2016 *River Publishers. All rights reserved.*

1 Introduction

Nowadays, Internet is a versatile facility for satisfying the people with various services related to various different fields. In this age of advancement of technologies, almost everything is now available over the internet that can help us in completing many tasks easily and conveniently with just few clicks whether it is any task of daily usage or some specific service which requires lot of research to be done beforehand. We can purchase anything and pay our bills online by going through various websites and choosing among variety of options. As a result, there might be some confidential data to purchase anything or pay online. So security should be taken into account. There are some mechanisms trying to provide secured information on the internet such as HTTPS, SSL etc. HTTPS was introduced to be used as a secure channel rather than HTTP connections. It provides reliable communication over the internet in terms of security. Almost every website today uses this protocol to run their business. However, one drawback of HTTPS is that it cannot tolerate SSL man in the middle attack that leads to security issue when the confidential information is hacked. Many researchers are working in this area to protect secure data by using cryptographic techniques, web browser plug-in and many more software's used to mitigate these attacks.

When a device needs to communicate with any other device on the same wireless network, it checks its ARP cache to find the MAC Address of the destination device. As a result of this check, if the MAC address is found in the cache, it is used for communication. If not found in local cache, the source machine generates an ARP request. The source broadcasts this request message to the local network. The message is received by each device on the LAN as a broadcast. ARP is a stateless protocol; therefore all client operating systems update their cache if a reply is received, inconsiderate of whether they have sent an actual or faked request. Since ARP does not offer any method for authenticating replies in the network, these replies are vulnerable to be manipulated by other hosts on a network. [3]

This paper provides the comparative analysis among the various defense strategies used to detect and mitigate these attacks based on the network para- meters time and scalability. It takes into account the experimental observations of each method and highlights the best method to mitigate these attacks.

The remainder section of this paper is organized as follows: Section 2 describes SSL MITM Background, Section 3 describes the experimenting details, Section 4 describes the comparative analysis and Section 5 concludes the paper.

2 SSL MITM Background

Secure Socket Layer/Transport Layer (SSL/TLS) [6] was originally developed by Netscape. It is used to deploy confidentiality, authenticity and integrity between web client and web server. SSL follow a standard handshake procedure to establish communication between client and server. The handshake prior to an HTTPS session follows:

1. The client contacts a server that hosts a secured URL.
2. The server responds to the client's request and sends the server's digital certificate (X.509) to the browser.
3. The client now verifies that the certificate is valid and correct. Certificates are issued by well-known authorities (e.g. Thawte or Verisign).
4. The server could optionally choose to confirm a user's identity. The TLS-enabled server software can check that the client's certificate is authorized and has been issued by a certificate authority (CA) from the server's list of trusted CAs. This confirmation is important in confidentiality as if the server is a bank sending confidential financial information to a customer and wants to check the recipient's identity.
5. Once the certificate is validated, random one-time session key is generated by the client to encrypt all communication with the server.
6. The client now encrypts the session key with the server's public key, which was transferred with the digital certificate. Encrypting using the server's public key assures that others cannot eavesdrop on this sensitive exchange.

In this way a secure session is established between the client and server, both knows the session key and can communicate via a secure channel.

However with the new intrusion tools developed, the SSL can be stripped apart. Moxie Marlinspike [9], during a BlackHat conference in 2009 released a tool known as SSLStrip which rendered the use of SSL digital certificates unfruitful. Here, the attacker created a fake digital certificate with spoofed identity and echoed the data both ways between client and the server. SSLStrip rewrites all HTTPS addresses as HTTP addresses and then saves traffic content. But these can have threatening effects on the organizations.

Security measures need to be taken to safeguard the interests of users. Various mechanisms that are used to mitigate these attacks are described in Section 3.

3 Defence Strategies Experimenting Details

One of the simplest ways to detect ARP poisoning is to list the ARP table which contains all the MAC addresses our computer knows. Now under ARP poisoning, there should necessarily be a duplicate MAC address. To detect it, open CMD as administrator and type the command:

"arp –d –a 3"

This command will clear the ARP cache i.e. it will clear possible disconnected devices on our network as shown in Figure 1.

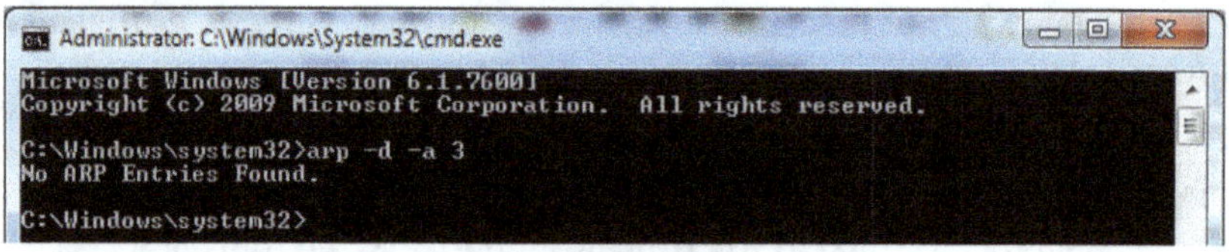

Figure 1   Clear ARP cache.

Now type the command:

"arp –a"

This command will list the ARP table and if there is any duplicate MAC address found in the ARP table. This is a clear indication of ARP poisoning.

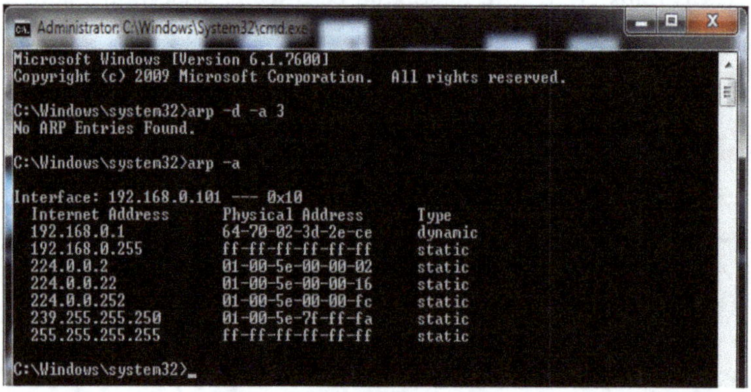

Figure 2   ARP table.

The disadvantage of this method is that it is not scalable for large networks as it would be very difficult to check each entry manually for duplicates.

Another method that is used to detect ARP poisoning is using Wireshark [7]. For the Wireshark to detect duplicate IP addresses, we need to capture the network traffic by selecting our network adapter from the interfaces list and click on start button to capture packets.

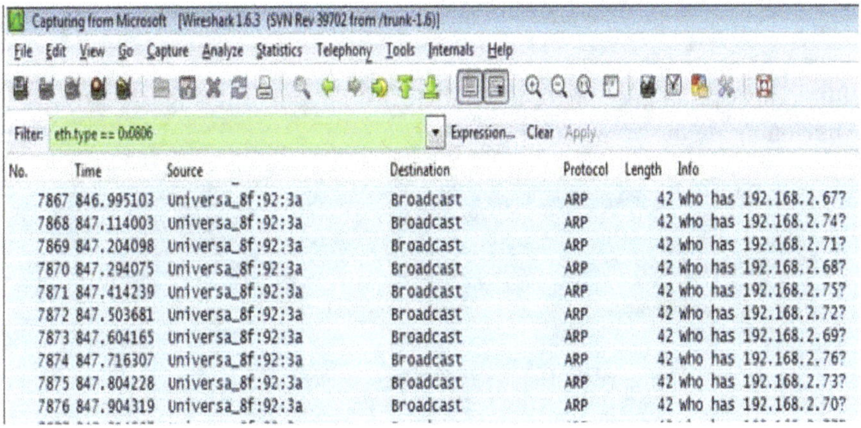

Figure 3   Normal traffic flow.

Now click on "edit preferences" from the top menu on right hand side as shown in Figure 4.

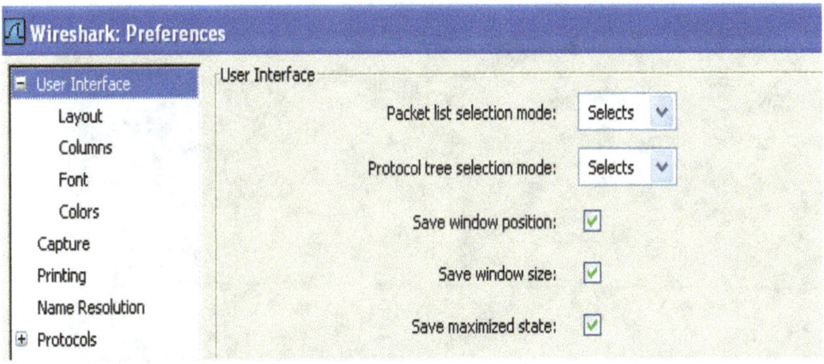

Figure 4   Edit preferences.

Select the protocol ARP/RARP from the Protocols tab and enable the feature "Detect duplicate IP address configuration" as in Figure 5.

Detect ARP request steems:	☑
Number of requests to detect during period:	30
Detection period (in ms):	100
Detect duplicate IP address configuration:	☑

Figure 5   Enabling detection.

When this option is enabled, it will detect if anyone is trying to perform ARP poisoning on the network.

```
9264 1033.334261 Universa_8f:92:3a       Azurewav_31:ff:65   ARP    42 192.168.2.1 is at 70:f3:95:8f:92:3a
9265 1035.335990 Universa_8f:92:3a       Azurewav_31:ff:65   ARP    42 192.168.2.1 is at 70:f3:95:8f:92:3a
9266 1037.337842 Universa_8f:92:3a       Azurewav_31:ff:65   ARP    42 192.168.2.1 is at 70:f3:95:8f:92:3a
⊞ Frame 9253: 42 bytes on wire (336 bits), 42 bytes captured (336 bits)
⊞ Ethernet II, Src: Universa_8f:92:3a (70:f3:95:8f:92:3a), Dst: Azurewav_31:ff:65 (00:15:af:31:ff:65)
⊞ [Duplicate IP address detected for 192.168.2.1 (70:f3:95:8f:92:3a) - also in use by 00:0e:2e:60:b7:22 (frame 8661)]
⊞ Address Resolution Protocol (reply)
```

Figure 6   Duplicate IP address detected.

In this way Wireshark can be used to detect ARP poisoning on the network.

Ettercap [5] supports a variety of loadable modules at run time known as plug-ins. These plug-ins are automatically compiled until we specify disable plug-in option to the configure script. The available plug-ins in Ettercap is shown in Figure 7.

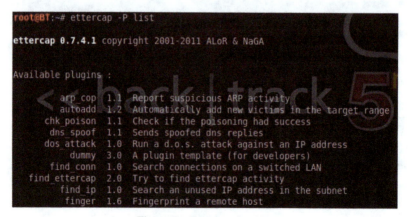

```
root@BT:~# ettercap -P list

ettercap 0.7.4.1 copyright 2001-2011 ALoR & NaGA


Available plugins :

         arp_cop  1.1  Report suspicious ARP activity
         autoadd  1.2  Automatically add new victims in the target range
      chk_poison  1.1  Check if the poisoning had success
       dns_spoof  1.1  Sends spoofed dns replies
      dos_attack  1.0  Run a d.o.s. attack against an IP address
           dummy  3.0  A plugin template (for developers)
       find_conn  1.0  Search connections on a switched LAN
   find_ettercap  2.0  Try to find ettercap activity
         find_ip  1.0  Search an unused IP address in the subnet
          finger  1.6  Fingerprint a remote host
```

Figure 7   Ettercap plug-ins.

There is one plug-in "find_ettercap" that is used to find the Ettercap activity.

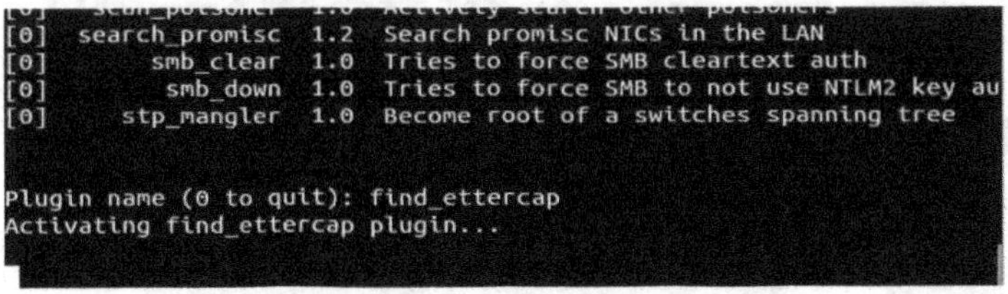

Figure 8 Find Ettercap activity.

The plug-in "search_promisc" is used to find if any machine is running in promiscuous mode.

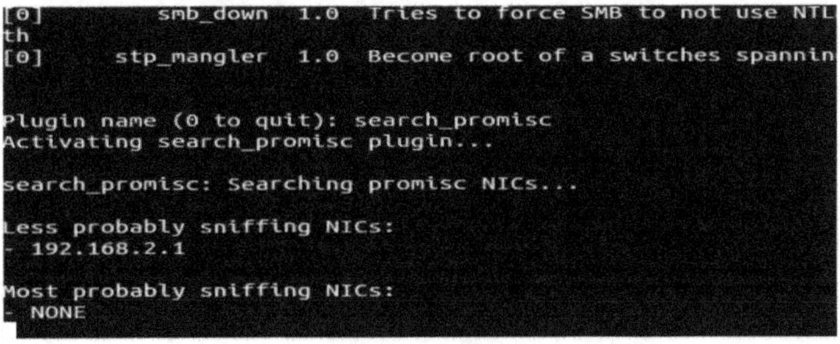

Figure 9 Searching promiscuous NIC.

Finally the "chk_poison" plug-in is used to check if anyone is trying to perform ARP poisoning on the network.

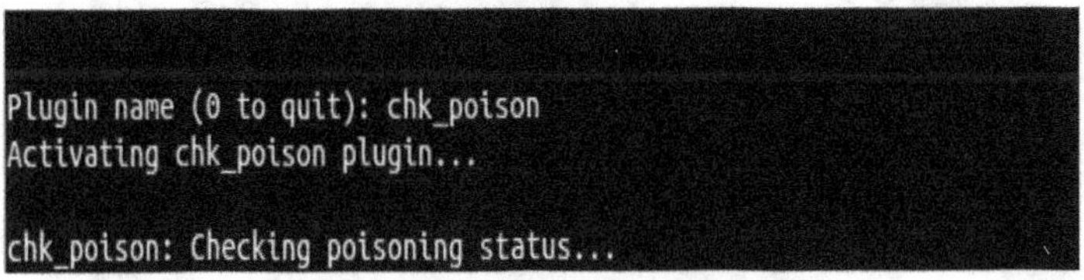

Figure 10 Checking ARP poisoning.

Intrusion Detection Systems like Snort [11] are available freely to detect any kind of attack. Snort is an open source system by Sourcefire. It is most widely deployed technology worldwide. Snort service is available in Backtrack. We need to start this service to monitor to start detection activity on the network.

Figure 11   Start snort.

We need to configure snort before using it. To configure it, we use vim editor and open the snort configuration file by the following command:

Figure 12   Opening snort configuration file.

In the Snort configuration file, we will see a line "var HOME_NET any". Change the word any to our own IP address as in Figure 13.

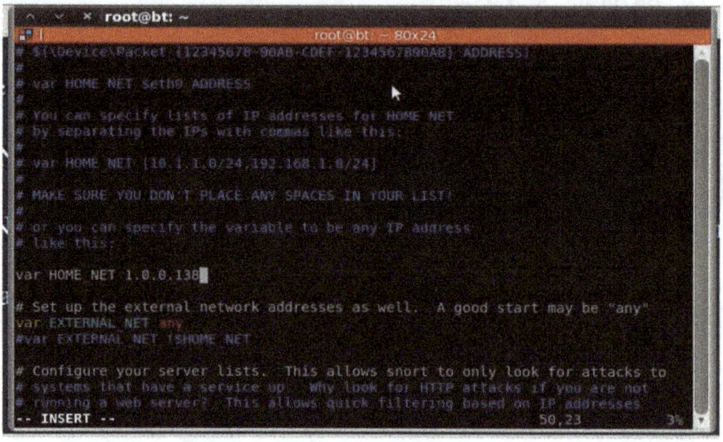

Figure 13   Configuring snort.conf file.

Now if anyone will try to perform any of kind of attack like man-in-the-middle attack, denial-of-service attack etc. Snort will detect it like in Figure 14.

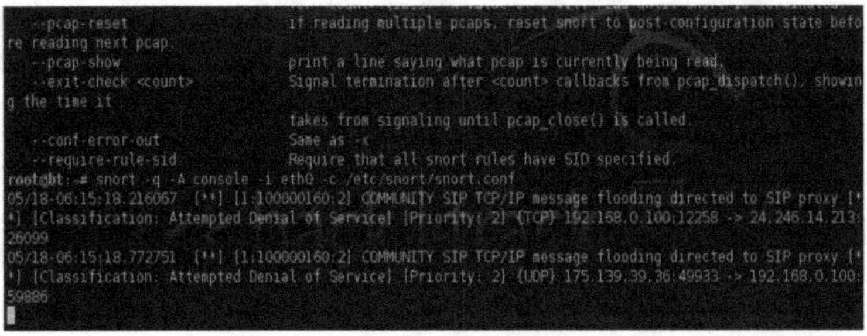

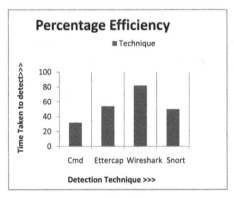

Figure 14 Denial of service attack detected.

In this way Snort can be used on Backtrack to detect any kind of attack.

4 Results and Analysis

To highlight the current state of art in this area, the most important and apt techniques have been tabulated in Table 1. The analysis reveals that if time is an important factor taken into consideration then detection using command prompt is least suited method to combat ARP poisoning assaults followed by Snort, then Ettercap plug-ins and then best method i.e. Wireshark. To support the view, a graphical analysis of various explored techniques has been given in Figure 15.

Figure 15 Bar graph analysis taking time into consideration.

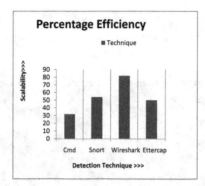

Figure 16   Bar Graph analysis taking scalability into consideration.

If scalability is an important factor taken into consideration then detection using command prompt is least suited method to combat ARP poisoning assaults followed by Snort, then Ettercap plug-ins and then best method i.e. Wireshark. To support the view, a graphical analysis of various explored techniques has been given in Figure 16.

5 Conclusion

It can be concluded that computer networks are vulnerable to dangerous attacks even today. Although every website today has preliminary protection on the information using HTTPS protocol instead of HTTP, still hacker is able to harvest the confidential information by using SSLstrip. He has to hook up to the same network with a victim and can monitor everything happening on the network. This paper as a result provided the analysis of best techniques to detect these attacks taking into consideration two network parameters time and scalability. The bar graph analysis shows that in both the cases Wireshark is the best method to deal with the attacks. In future, an ideal mechanism can also be developed and implemented by accomplishing several experiments with reference to the work done in wide scenarios of ARP spoofing attacks.

References

[1] "ARP-Guard," (accessed 28-July-2013). [Online]. Available: http://www.arp-guard.com.

[2] B. D. Schuymer, "ebtables: Ethernet bridge/switch tables," Mar. 2006, (accessed 28-July-2013). Available:http://ebtables.sourceforge.net.

[3] D. Plummer. An ethernet address resolution protocol, Nov. 2010. RFC 826.

[4] http://www.backtrack-linux.org/

[5] http://ettercap.sourceforge.net/

[6] https://www.digicert.com/ssl.htm

[7] https://www.wireshark.org/

[8] L. N. R. Group. arpwatch, the Ethernet monitor program; for keeping track of ethernet/ip address pairings. (Last accessed April 17, 2012).

[9] Moxie Marlinspike, "SSLStrip, Black Hat DC 2009", Retrieved http://www.thoughtcrime.org/software/sslstrip/

[10] S. Whalen, "An introduction to ARP spoofing," 2600: The Hacker Quarterly, vol. 18, no. 3, Fall 2001,. Available:http://servv89pn0aj.sn.sourcedns.com/_g bpprorg/ 2600/arp spoofing intro.pdf

[11] Snort Project, The Snort: The open source network intrusion detection system. <http://www.snort.org>.

[12] Demuth and A. Leitner. ARP spoofing and poisoning: Traffic tricks. *Linux Magazine*, 56:26–31, July 2011.

Information Security Risk Assessment of Smartphones Using Bayesian Networks

Kristian Herland, Heikki Hämmäinen and Pekka Kekolahti

Aalto University, School of Electrical Engineering,
Department of Communications and Networking, Espoo, Finland
Email: {kristian.herland; heikki.hammainen; pekka.kekolahti}@aalto.fi

Received 31 August 2015; Accepted 20 November 2015;
Publication 22 January 2016

Abstract

This study comprises an information security risk assessment of smartphone use in Finland using Bayesian networks. The primary research method is a knowledge-based approach to build a causal Bayesian network model of information security risks and consequences. The risks, consequences, probabilities and impacts are identified from domain experts in a 2-stage interview process with 8 experts as well as from existing research and statistics. This information is then used to construct a Bayesian network model which lends itself to different use cases such as sensitivity and scenario analysis. The identified risks' probabilities follow a long tail wherein the most probable risks include *unintentional data disclosure, failures of device or network, shoulder surfing or eavesdropping* and *loss or theft of device*. Experts believe that almost 50% of users share more information to other parties through their smartphones than they acknowledge or would be willing to share. This study contains several implications for consumers as well as indicates a clear need for increasing security awareness among smartphone users.

1 Introduction

The global number of smartphone users already surpassed 1 billion in 2012 [18] and in Finland, the share of smartphones relative to all mobile handsets in use exceeded 50% in 2013 [26]. As smartphones have become powerful

Journal of Cyber Security, Vol. 4, 153–174.
doi: 10.13052/jcsm2245-1439.424
ⓒ 2016 *River Publishers. All rights reserved.*

enough to fulfil most consumers' computing needs, users are effectively migrating their computing tasks from traditional computers to smartphones. While traditional computer security is common knowledge and end-users typically use security software such as anti-virus on these devices, smartphone security is not as well understood among end-users.

Research concerning specific smartphone vulnerabilities exists in large numbers. Terms such as mobile malware and mobile phishing already return numerous matches in research paper searches. However, comprehensive risk assessments of smartphone use are not readily available. It is not immediately clear how much mobile malware contributes to the information security breaches that occur via smartphones, for example. Moreover, it is unclear how much smartphone use contributes to all information security breaches.

The main objective of this study is to perform a high-level risk assessment of information security related to smartphone usage. As a secondary objective, this study aims to design and implement a practical risk assessment process for eliciting information from multiple experts and consolidating this information into a Bayesian network. The outcome of this risk assessment is a Bayesian network model of information security risks, which can be used for various purposes such as scenario and sensitivity analysis.

2 Bayesian Networks

First documented applications of Bayesian networks in risk analysis include evaluation of terrorism threats by Hudson et al. in 2001 [5] and of structures under fire by Gulvanessian and Holicky in 2002 [4]. Later, Bayesian networks have been used for risk management purposes in various fields such as banking [7], nuclear power plants [6], building fires [10], earthquakes [22] and other natural hazards [23]. Some studies employ fully knowledge-based approaches using only expert interview [6, 8, 9], whereas others combine expert opinion with statistical knowledge [7, 37, 38]. Eunchang et al. [24] describe a study where 252 industry experts were surveyed for risk knowledge concerning a large engineering project. Bayesian network models have also been used for assessing the probability of ship collision [37] and effectiveness of oil combating [38] in the Gulf of Finland.

Peltola and Kekolahti [30] use Bayesian networks and expert knowledge to perform a risk assessment of the Finnish TETRA PSS network, where several risk sources and controls affect the availability of the network. Bayesian networks have been also used in several research papers to model interdependent information security vulnerabilities as attack scenarios

[13–15, 17]. Sommestad et al. [16] present a framework for analysing cyber security using Bayesian statistics and Mo et al. [12] propose a quantitative model for evaluating a firm's cyber security readiness by use of Bayesian networks.

Classical methods of causal and frequency analysis include, in addition to Bayesian networks, Fault Trees [32], Markov chains [33] and Petri nets [34]. Fault trees are commonly used in causal analysis but can be completely replaced by Bayesian networks. Markov chains and Petri nets on the other hand are not suitable for causal analysis [21]. Bayesian networks was chosen in this study as the risk analysis method due to the following reasons:

1. Efficient consolidation of hard data and expert opinion [35].
2. Ability to capture causal knowledge even from domain experts with little or no statistical experience [2].
3. Easily understandable format for visualizing causal relationships between variables [1].
4. Suitability for simple expert elicitation methods [3].
5. Robustness with regards to incomplete information [35].
6. Flexibility and abundance of use cases [3].

However, Bayesian networks also exhibit the following disadvantages:

1. Continuous variables typically must be discretized before use [21]. However, the variables analysed in this study are ordinal and therefore, this disadvantage is irrelevant.
2. Determining quantitative values via expert elicitation is a complex [35] and time-consuming process [1]. However, the number of variables to be elicited can be reduced using for example parameterized methods. A secondary target of this study is to simplify the elicitation by developing an interview process supported by an Excel-based tool for gathering data and deriving Node Probability Tables (NPT).

3 Research Method

Whereas in the data-driven approach the model structure is first learned using either score- or constraint-based methods [39–41] and thereafter the parameters by learning the local distributions implied by the structure, the Bayesian network model in this study is constructed using a knowledge based approach due to the lack of available data. First, relevant a priori information is reviewed from literature in order to determine the known assets and risks related to smartphone use. This information is then utilized as a basis for interviews with domain experts, where information is gathered in order to construct a Bayesian network model of the risks and consequences. This model is then used for further analysis of the risks.

The expert interviews follow a two-stage expert elicitation process designed in this study. This process aims to collect the data necessary for constructing a Bayesian network model of information security risks and assets related to smartphone use in Finland at the time of this study. The purpose of the first stage is to gather enough information to build a qualitative model of the information security risks and consequences, i.e. the graphical BN structure in which nodes represent risks or consequences, and edges indicate causal relationships. During the second stage, this model is presented and validated with each expert whereafter the strengths of dependencies and impacts are determined, i.e. the quantitative values are elicited. Figure 1 describes this process.

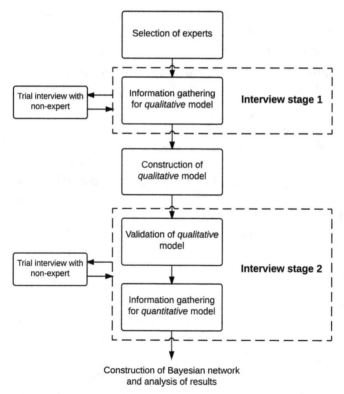

Figure 1   High level overview of expert interview process.

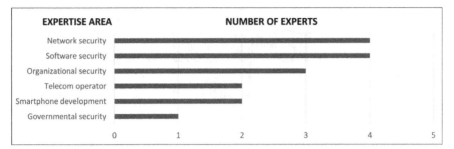

Figure 2 Interviewed experts' areas of expertise.

The experts to be interviewed are chosen based on experience, subject matter knowledge, current employer and position. Two main objectives are taken into account when choosing the experts: (1) the group of experts should include various specializations with partial overlap in order to ensure completeness of the information available and (2) the interviewee's experience and knowledge in the domain and their own viewpoint should be good or very good in order to ensure the quality of the information available. The representation of different expertise areas within the chosen experts is visualized in Figure 2.

3.1 Interview Stage 1

Two different types of techniques are commonly used to capture the information required for building the qualitative Bayesian network from expert interviews. *Structured* techniques involve specific questions about predefined concepts and are thus most suitable for confirming existing knowledge. *Unstructured* approaches focus on exploring new information and are thus well suited for use in domains for which existing knowledge is lacking or non-existent. Prior knowledge in smartphone security is available but in limited extent and due to this lack of completeness, a combination of *structured* and *unstructured* methods are used. The first stage interviews roughly follow the process visualized in Figure 3.

3.2 Interview Stage 2

The main objective of the second stage interviews is to collect the information necessary in order to construct the quantitative Bayesian network model, which is the set of Node Probability Tables (NPT) assigned to the nodes of the

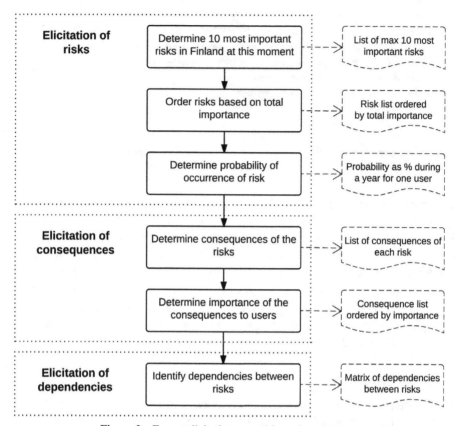

Figure 3   Expert elicitation stage 1 interview process.

qualitative Bayesian network. In this model, each risk is defined as a Boolean node and each consequence as a ranked node with the scale *negligible-low-medium-high*.

Eliciting the content of NPT's from experts can be performed in several ways. Manual elicitation by interview quickly becomes infeasible in non-trivial networks due to the exponential increase in NPT size with the amount of nodes. A less time-consuming alternative is utilizing a parameterized model, where the NPTs are constructed according to a formula whose variables are elicited from experts. Common methods of building parameterized models include using NoisyOR operators [27, 28], their multivalued generalization Noisy-MAX or a truncated normal distribution [25], for instance. However, these methods are not suitable for the purpose of this study due to their prerequisites, i.e. that either all nodes are Boolean, represent abstractions of continuous variables or have only few parent nodes.

For the purpose of this risk assessment, a method is designed with the objective of being easy to understand even by experts with little or no statistical experience. The resulting method consists of assessing each risk-consequence-pair individually and thereafter combining this information into NPTs using an Excel-based tool. The process is visualized in Figure 4.

This method was easy for experts to understand and follow, minimizing the time and effort required for familiarizing the experts with the process. However, due to the high amount of risk-consequence-pairs, this method was more time consuming than typical parameterized methods. On the other hand, this method is not as prone to the typical loss of accuracy observed when representing large amounts of variables with a simplified function. Compared to manual elicitation of NPTs, this method still provided a nearly hundredfold reduction in variables to be elicited.

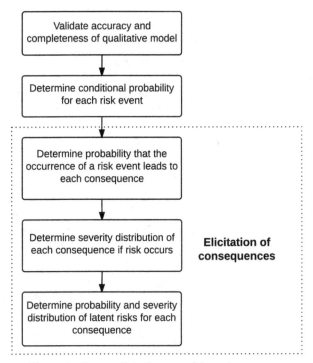

Figure 4 Expert elicitation stage 2 interview process.

4 Results

The Bayesian network constructed in this study describes the most important risk events and consequences related to smartphone use in Finland during year 2015. Probabilities in the network describe the expected probability of an event or consequence for a single smartphone user during one year without any additional knowledge of the user. The probabilities and severity distributions are arithmetic averages of the values given by experts. The choice of identified risks to be included in the network was made based on the interviewed experts' opinions of the risks' importance, taking into account both probability and consequences.

A subset of the complete model be seen in Figure 5, which describes the risks that could cause either the consequence *leakage of personal* or *leakage of confidential data*. This qualitative subset describes the most important risk events identified by the experts but does not include quantitative values such as probabilities of risk events. The qualitative model subsets, such as shown in Figure 5, were used for eliciting quantitative values from the experts during interview stage 2.

Figure 6 shows the complete Bayesian Network structure including the nodes' state probability distributions. For example the risk *shoulder surfing or eavesdropping* can directly cause leakage of confidential or personal data as well as lead to theft of the device, for example in a scenario where a thief steals the device after seeing its passcode. Furthermore, the thief might be interested in accessing the device's information and services, thus realizing the risk *unauthorized physical device access*, as opposed to wiping the device's memory and selling it.

Figure 7 visualizes the probability of occurrence for each risk, which shows that most risks are unlikely to occur during one year. However, based on Figure 7, experts believe that almost 50% of smartphone users become victim to *unintentional data disclosure* during a year, i.e. share more information to other parties through their smartphones than they acknowledge or would be willing to share.

Figure 8 visualizes the possible consequences of each risk. The total bar length indicates each consequence's probability when the respective risk event occurs. In addition, the bars are divided by colour into different consequence severities, wherein each severity has been assigned a probability. Consequences with a negligible probability or severity are omitted from the figure. Based on the figure, the risks *unintentional data disclosure* and *vendor backdoor* are both very likely to cause *leakage of personal* data when they occur but the effects of *vendor backdoor* are more likely to be severe.

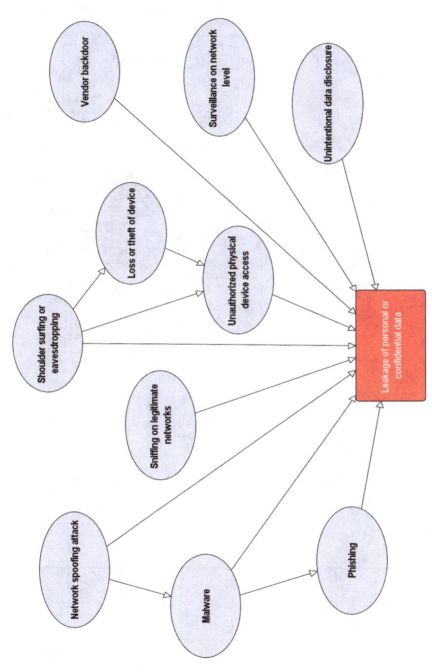

Figure 5 Qualitative model of the risk events which can cause the consequences *leakage of personal data* or *leakage of confidential data*.

162   *K. Herland et al.*

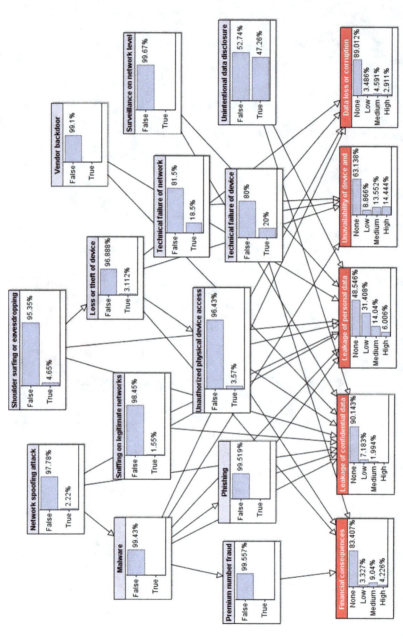

Figure 6   Complete Bayesian network model of information security risk events related to smartphone use, and their respective consequences.

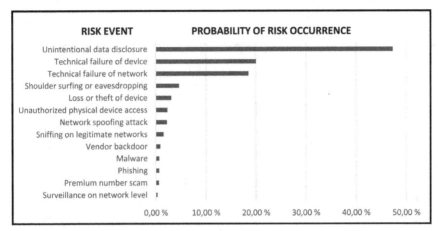

Figure 7 Probability of occurrence for each risk.

From the risk assessment results, it seems clear that a data breach is considerably more likely to concern a smartphone user's personal information than confidential information related to the user's employer. One likely reason is that while businesses and governmental entities have significant incentives to use encryption and strong authentication, consumers' often value price and ease of use more than security.

Figure 9 shows a sensitivity graph describing the effects of individual risk events on medium- or high-severity confidential data leakage. According to the analysis, the consequence is most sensitive to *unauthorized physical device access* or *malware*. This is reasonable as an unauthorized user would have access to all services which do not require additional authentication and malware with elevated rights could access all data on the device and monitor interaction between the device and user. However, most devices used for confidential purposes should require a passcode for unlocking the device, which might not be sufficiently represented in the results.

Figure 10 represents the effects of individual risks on medium- or high-severity personal data leakage. The results resemble those of confidential data leakage in Figure 9. However, two clear differences exist between these results: (1) all risk events have a significantly higher probability of affecting personal data than confidential data and (2) the effect of *unintentional data disclosure* is much higher relative to other risks' effects on leakage of personal than confidential data.

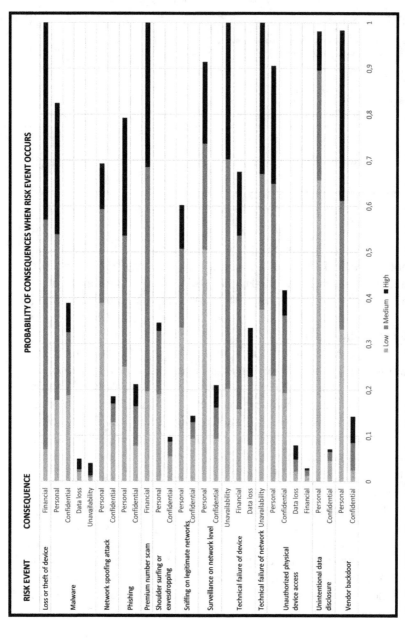

Figure 8 Probability of each risk event's consequences and consequence severities (low-medium-high), when the respective risk event occurs. Risk events ordered alphabetically.

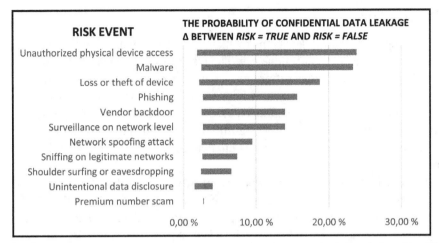

Figure 9 Effect of individual risk events on medium- or high-severity leakage of confidential data.

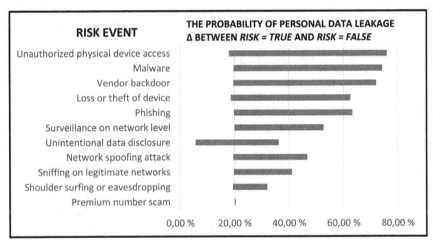

Figure 10 Effect of individual risk events on medium- or high-severity leakage of personal data.

Loss or corruption of data related to smartphone usage seems to be a relatively unlikely scenario and most often caused by *loss or theft of device* or *technical failure of device*. Also, low-severity and medium-severity unavailability is very rarely caused by anything else than technical failures of the device or network. On the other hand, high unavailability is rarely caused by anything else than *loss or theft the device*.

5 Discussion

The Bayesian network model seems realistic as a representation of the actual risk space surrounding smartphone use. The results are logical and offer more insight into the importance of different risks and their consequences. Bayesian networks as a method seems suitable for information security risk assessment as it produces a very flexible model that can be used for various kinds of analysis. However, the process for eliciting information and constructing the network is time-consuming and poses challenges for networks with a large amount of nodes.

The information elicited from experts is mostly well in line with that found in statistics and existing research. For example, the conditional probability that the data stored on a lost or stolen smartphone is accessed, reflects the results of Symantec's Honey Stick Project [36]. The experts' answers also varied significantly for some risks, such as the unintentional data disclosure risk, where estimates of probability ranged from less than 1% to 95%. However, the average of answers received from the 8 experts seem realistic. One special threat that arose in this risk assessment is the speculative vendor backdoor, for which the experts had very different opinions. According to some experts, at least 10% of the smartphones in Finland are likely to include an active backdoor designed by the device vendor or operating system developer, whereas some experts believe that none of the smartphones used in Finland have such a backdoor. Experts' answers were often derived from a real-world assumption, such as that exactly one mid-sized smartphone vendor is inserting backdoors into mobile devices.

Some experts were uncomfortable estimating probabilities without any statistical data. For instance data leaks regarding company confidential information is a topic where experts felt particularly insecure due to lack of hard data, especially as businesses do not necessarily report all data breaches to outside parties due to the possibility of reputational damage.

A common expectation between experts is that the mobile risk space is going to undergo significant changes in the near future. Expectations include increased usage of mobile banking and payment services as well as a respective increase in attacks that target these services. Also, experts expect users to increasingly store sensitive information on their smartphones. The abundance of sensor data on smartphones is still a relatively unexploited resource, for which risks such as *malware* and *vendor backdoor* present a natural threat. Research shows that many mobile services are vulnerable to sophisticated and unsuspicious phishing attacks [29].

The Bayesian network model can be used as is to illustrate the significance of different smartphone risks in different scenarios. Based on this study, the following implications for consumers can be drawn. First, consumers should ensure that their devices' built-in security features are enabled, especially a passcode and restrictions against installing applications from outside the official application store. Second, consumers should familiarize themselves with mobile applications' privacy settings and terms. Third, consumers should regard their smartphones as untrusted devices, wherein a minimal amount of private or confidential information should be stored on them. Also, it is warranted for security stakeholders to place more emphasis on educating users about smartphone security features and pitfalls.

The model could be extended with controls and mitigants, which contribute to the probability or consequences of risk events. Such an extended model could be used for identifying the most dangerous practices of smartphone usage as well as which security measures would be most useful for prevention or mitigation of risks. Further extended with costs of security measures, the model would lend itself to cost-benefit analysis of security measures. Another possible extension of the network would be to add demographic parameters such as age, occupation or income of the user, wherein the network would give more insight into the vulnerabilities and security awareness of each demographic group. The model constructed in this study could be applied to other countries by updating the parameter values and structure as necessary to represent the local environment. In order to ensure validity of the model in any environment, the parameter values as well as the structure should be updated regularly.

6 Conclusion

The purpose of this study is to perform an information security risk assessment of smartphones using Bayesian networks. Most information is gathered during a two-stage expert elicitation process, where the experts represent various experience, knowledge and viewpoints related to information security, thus ensuring the completeness of the elicited information. The expert interviews follow a process designed as part of this study in order to facilitate accurate and simple elicitation.

The outcome of this study is a Bayesian network model which documents the information security risks related to smartphone use and can be extended with new data when available. The model shows that the most important risks in Finland include traditional information security risks such as *malware* and *phishing*, very general risks such as *loss or theft of device* and relatively new risks such as *unintentional data disclosure* through legitimate applications. In fact experts believe that almost 50% of smartphone users share more information to other parties through their smartphones than they acknowledge or would be willing to share. Also, the experts raise a concern over more speculative risks such as *surveillance on network level* and *vendor backdoors*. The study contains several implications to consumers as well as highlights a need for increasing security awareness among smartphone users.

Bayesian networks are found to be an effective method for documenting and analysing causal knowledge of domain experts. The model lends itself well to different types of sensitivity analysis, which would be especially useful when analysing potential controls and mitigants for risks. The expert elicitation method designed was easy for experts to understand and delivered accurate results, but however time-consuming. Both Bayesian networks and the expert elicitation method could be applied to other risk assessments as well.

Future research topics include extending the model with controls and mitigants in order to identify causes and preventive measures or with demographic parameters to better identify how risks vary between different age, occupation, income or location groups. Further research is also warranted for developing more effective tools and methods for expert elicitation and consolidation of results to be used in Bayesian networks.

References

[1] Fenton, N., Neil, M. *Risk Assessment and Decision Analysis with Bayesian Networks*. Noca Raton: CRC Press, 2013.

[2] Nadkarni, S., Shenoy, P. *A causal mapping approach to constructing Bayesian networks*. Decision Support Systems, 2004, volume 38, p. 259–181.

[3] Weber, P., Medina-Oliva, G., Simon, C., Iung, B. *Overview on Bayesian networks applications for dependability, risk analysis and maintenance areas*. Engineering Applications of Artificial Intelligence, 2012, volume 25 (4), p. 671682.

[4] Gulvanessian, H., Holicky, M. *Determination of actions due to fire: recent developments in Bayesian risk assessment of structures under fire*. Progress in Structural Engineering and Materials, 2002, volume 3 (4), p. 346–352.

[5] Hudson, L., Ware, B., Laskey, K., Mahoney, S. *An Application of Bayesian Networks to Antiterrorism Risk Management for Military*

Planners, 2002. [Online] Available from: http://www.mathcs.emory.edu/~whalen/Papers/BNs/KathyLanskey/Antiterrorism.pdf [Accessed 4 March 2015]

[6] Kim, M., Seong, P. *A computational method for probabilistic safety assessment of I&C systems and human operators in nuclear power plants*. Reliability Engineering & System Safety, 2006, volume 91 (5), p. 580–593.

[7] Cornalba, C., Giudici, P. *Statistical models for operational risk management*. Physica A: Statistical Mechanics and its Applications, 2004, volume 338 (1–2), p. 166–172.

[8] Russel A., Quigley J., Van der Meer R. *Modelling the reliability of search and rescue operations with Bayesian Belief Networks*. Reliability Engineering & System Safety, 2008, volume 93 (7), p. 940–949.

[9] Trucco P., Cagno E., Ruggeri F., Grande O. *A Bayesian Belief Network modelling of organisational factors en risk analysis: A case study in maritime transportation*. Reliability Engineering & System Safety, 2008, volume 93 (6), p. 845–856.

[10] Hanea D., Ale B. *Risk of human fatality in building fires: A decision tool using Bayesian networks*. Fire Safety Journal, 2009, volume 44 (5), p. 704–710.

[11] Cheon S-P., Kim S., Lee S-Y., Lee, C-B. *Bayesian networks based rare event prediction with sensor data*. Knowledge-Based Systems, 2009, volume 22 (5), p. 336–343.

[12] Mo, S. Beling, P. Member, Crowther, K. Quantitative Assessment of Cyber Security Risk using Bayesian Network-based model. In *Systems and Information Engineering Design Symposium*, Charlottesville, VA, 2009, p. 183–187.

[13] Noel, S., Jajodia, S., Wang, L., Singhal, A. *Measuring Security Risk of Networks Using Attack Graphs*. International Journal of Next Generation Computing, 2010, volume 1 (1), p. 1–11.

[14] Khosravi-Farmad, M., Rezace, R., Harati, A., Bafghi, A. Network Security Risk Mitigation Using Bayesian Decision Networks. In *4th International eConference on Computer and Knowledge Engineering (ICCKE)*, Mashhad, Iran, 2014, p. 267–272.

[15] Dantu, R., Kolan, P. *Risk Management Using Behavior Based Bayesian Networks*. Intelligence and Security Informatics, 2005, volume 3495, p. 115–126.

[16] Sommestad, T., Ekstedt, M., Johnson, P. Cyber Security Risks Assessment with Bayesian Defense Graphs and Architectural Models. In *42nd*

Hawaii International Conference on System Sciences, Big Island, HI, USA, 2009, p. 1–10.

[17] Cie, P., Li, J., Ou, X., Liu, P., Levy, R. Using Bayesian Networks for Cyber Security Analysis. In *IEEE/IFIP International Conference on Dependable Systems and Networks (DSN)*, Chicago, IL, USA, 2010, 211–220.

[18] Strategy Analytics. *Worldwide Smartphone Population Tops 1 Billion in Q3 2012*, 17 Oct 2012. [Online] Available from: http://www.businesswi re.com/news/home/20121017005479/en/StrategyAnalytics-Worldwide-Smartphone-Population-Tops-1 [Accessed 4 March 2015]

[19] Strategy Analytics. *Global Mobile Phone Shipments Reach 460 Million Units in Q3 2014*, 30 Oct 2014. [Online] Available from: http://blogs.strategyanalytics.com/WDS/post/2014/10/30/ StrategyAnalytics-Global-Mobile-Phone-Shipments-Reach-460-Million -Unitsin-Q3-2014.aspx [Accessed 4 March 2015]

[20] Omlis, *Global Mobile Payment Snapshot 2014*, 5 Aug 2014. [Online] Available from: http://www.omlis.com/omlis-media-room/worldwide-use-ofmobile-payments/ [Accessed 4 March 2015]

[21] Rausand, M. *Risk Assessment: Theory, Methods, and Applications.* New Jersey: Wiley, 2011.

[22] Bayraktarli Y., Ulfkjaer J., Yazgan U., Faber M. On the application of bayesian probabilistic networks for earthquake risk management. In *9th International Conference on Structural Safety and Reliability (ICOSSAR 05)*, Rome, Italy, 2005.

[23] Straub D. Natural hazards risk assessment using Bayesian networks. In *9th International Conference on Structural Safety and Reliability (ICOSSAR 05)*, Rome, Italy, 2005.

[24] Eunchang, L., Park, Y., Shin, J. *Large engineering project risk management using a Bayesian belief network.* Expert Systems with Applications, 2009, volume 36 (3), p. 5880–5887.

[25] Fenton, N., Neil, M., Caballero, J. *Using Ranked Nodes to Model Qualitative Judgments in Bayesian Networks.* IEEE Transactions on Knowledge and Data Engineering, 2007, volume 19 (10), p. 1420–1432.

[26] Vesselkov, A., Riikonen, A., Hämmäinen, H. *Mobile Handset Population in Finland 2005–2013*, Aalto University Department of Communications and Networking, 2014. [Online] Available from: https://research.comnet.aalto.fi/ public/Mobile_Handset_Population_200 5-2013.pdf [Accessed 5 April 2015]

[27] Huang, K., Henrion, M. Efficient Search-Based Inference for Noisy-OR BeliefNetworks. In *Twelfth Conference on Uncertainty in Artificial Intelligence*, Portland, OR, 1996, 325–331.

[28] Díez, F.J. Parameter adjustment in Bayes networks: the generalized noisy orgate. In *Ninth Conference on Uncertainty in Artificial Intelligence*, Washington D.C, 1993, 99–105.

[29] Felt, A., Wagner, D. *Phishing on Mobile Devices*, Workshop on Web Security and Privacy (W2SP), 2011. [Online] Available from: http://w2spconf.com/2011/papers/felt-mobilephishing.pdf [Accessed 1.7.2015]

[30] Peltola, M., Kekolahti, P. Risk Assessment of Public Safety and Security Mobile Service. In *International Conference on Availability, Reliability and Security ("ARES")*, Toulouse, France, 2015.

[31] Wang, J., Guo, M. Vulnerability Categorization Using Bayesian Networks. In *Proceedings of the Sixth Annual Workshop on Cyber Security and Information Intelligence Research*, Oak Ridge, Tennessee, USA, 2010, no. 29, p. 1–4.

[32] Fischhoff, B., Slovic, P., Lichtenstein, S. *Fault trees: Sensitivity of estimated failure probabilities to problem representation*. Journal of Experimental Psychology: Human Perception and Performance, 1978, volume 4(2), p. 330–344.

[33] Kemeny, J.G., Snell, J.L. *Finite markov chains*. Princeton, NJ: van Nostrand, 1960.

[34] Murata, T. *Petri nets: Properties, analysis and applications*. Proceedings of the IEEE, 1989, volume 77(4), p. 541–580.

[35] Uusitalo, L. *Advantages and challenges of Bayesian networks in environmental modelling*. Ecological Modelling, 2007, volume 203(3–4), p. 312–318.

[36] Symantec, *The Symantec Smartphone Honey Stick Project*, 2012. [Online] Available from: http://www.symantec.com/content/en/us/about/presskits/b-symantec-smartphone-honey-stick-project.en-us.pdf [Accessed 17.7.2015]

[37] Hänninen, M., Kujala, P. *Influences of variables on ship collision probability in a Bayesian belief network model*. Reliability Engineering & System Safety, 2012, volume 102, p. 27–40.

[38] Helle, I., Lecklin, T., Jolma, A., Kuikka, S. *Modeling the effectiveness of oil combating from an ecological perspective – A Bayesian network for the Gulf of Finland; the Baltic Sea*. Journal of Hazardous Materials, 2011, volume 185(1), p. 182–192.

[39] Singh, M., Valtorta, M. *Construction of Bayesian network structures from data.* International Journal of Approximate Reasoning, 1993, volume 12(2), p. 111–131. [Online] Available from: http://www.sciencedirect.com/science/article/pii/0888613X9400016V [Accessed 17.7.2015]

[40] Kjaerulff, U. B., Madsen, A. L. *Bayesian networks and influence diagrams.* New York: Springer, 2008.

[41] Scutari, M. *Learning Bayesian Networks with the bnlearn R Package.* Journal of Statistical Software, 2010, volume 35(3), p. 1–22.

Biographies

K. Herland is a cyber security specialist with an M.Sc. (Tech.) from the Department of Communications and Networking, Aalto University, Finland. He works as a security consultant for various public and private sector clients regarding security-related topics from technical IT security to organization-wide risk management. His special interests lie in the security of mobile devices and related technologies.

H. Hämmäinen is professor of networking technology at Department of Communications and Networking, Aalto University, Finland, since 2003. He received his Ph.D. in computer science from the same university in 1992. His main research interests are in techno-economics and regulation of mobile services and networks. Special topics recently include measurement and analysis of mobile Internet usage, value networks of cognitive radio, and diffusion of Internet protocols in mobile.

P. Kekolahti is a postgraduate student at the Department of Communications and Networking, Aalto University, Finland. His research interest is in the modeling of variety of complex telecommunications business related phenomena using Bayesian Networks. Pekka Kekolahti holds a M.Sc. and Lic.Sc.(Technology) from Helsinki University of Technology.

Digital Forensic Investigations: Issues of Intangibility, Complications and Inconsistencies in Cyber-Crimes

Ezer Osei Yeboah-Boateng[1] and Elvis Akwa-Bonsu[2]

[1]Ghana Technology University College (GTUC)
[2]Detectware Limited, Ghana
Email: eyeboah-boateng@gtuc.edu.gh; elvis@detect-ware.net

Received 31 August 2015; Accepted 20 November 2015;
Publication 22 January 2016

Abstract

The use of the Internet and computing resources as vital business tools continue to gain prominence day-by-day. Computing resources are utilized to create innovative and value-added products and services. Associated with this trend is the extent of cyber-crimes committed against or using computers. Experts anticipate that the extent and severity of cyber-attacks have increased in recent times and are likely to explode, unless some mitigation measures are instituted to curb the menace. As a response to the growth of cyber-crimes, the field of digital forensics has emerged.

Digital forensic investigations have evolved with the passage of time and it's impacted by many externalities. A number of key challenges ought to be addressed, such as the intangibility, complications and inconsistencies associated with the investigations and presentation of prosecutorial artefacts. The digital evidence is usually intangible in nature, such as an electronic pulse or magnetic charge. The question is how can the intangibility of computer crime complicate the digital forensic investigations? To what extent can inconsistencies during the investigation mar the permissibility or admissibility of the evidence?

Journal of Cyber Security, Vol. 4, 175–192.
doi: 10.13052/jcsm2245-1439.425
© 2016 *River Publishers. All rights reserved.*

This study is an experimentally exploratory set-up with virtual systems subjected to some malware exploits. Using live response tools, we collected data and analyzed the payloads and the infected systems. Utilizing triage information, memory and disk images were collected for analysis. We also carried out reverse engineering to decompose the payload.

The study unearthed the digital truth about malwares and cyber-criminal activities, whilst benchmarking with standard procedures for presenting court admissible digital evidence. The timelines of activities on infected systems were reconstructed. The study demonstrated that externalities of intangibility, complications and inconsistencies can easily mar digital forensic investigations or even bring the entire process to an abrupt end. Further studies would be carried out to demonstrate other ways perpetrators use in concealing valuable digital evidence in a cyber-crime.

Keywords: Digital Forensic Investigation, Cyber-crime, Digital evidence, Artefacts, Malwares, Payload.

1 Introduction

The use of the Internet and computing resources as vital business tools continue to gain prominence day-by-day. Computing resources are utilized to create innovative and value-added products and services [1]. Associated with this trend is the extent of cyber-crimes committed against or using computers. Experts anticipate that the extent and severity of cyber-attacks have increased in recent times and are likely to explode, unless some mitigation measures are instituted to curb the menace [2]. As a response to the growth of cyber-crimes, the field of digital forensics has emerged.

Typically, digital forensics involves carefully collecting and examining electronic evidence or artefacts, as well as accurate analysis and interpretation of collected evidence. This investigative process assesses the extent of damage to a compromised or an attacked system, as well as recovers lost information from such compromised system and ultimately, to present the digital evidence to prosecute the cyber-crime perpetrators. It has become imperative that law enforcement officers and digital forensics examiners adhere to high standards of the profession, if digital evidences were to be permissible in a competent court of jurisdiction.

Digital forensic investigations have evolved with the passage of time and its impacted by many externalities [3]. A number of key challenges ought to be addressed, such as the intangibility, complications and inconsistencies associated with the investigations and presentation of prosecutorial artefacts. The digital evidence is usually intangible in nature, such as an electronic pulse or magnetic charge. The question is, how can the intangibility of computer crime complicate the digital forensic investigations? To what extent can inconsistencies during the investigation mar the permissibility or admissibility of the evidence?

Cases abound whereby suspects were either incriminated or set free due to misinterpretation of digital evidence and/or inaccurate methods employed in collecting and analyzing the data.

The objectives of the study are:

- To collect and examine electronic evidence or artefacts – to assess the extent of damage and to recover lost information or data;
- To present digital evidence that would be admissible in court for the prosecution of cyber-crime perpetrators;
- To examine the externalities (network effects) that are likely to affect the admissibility of digital evidence in court, as well as to render the forensic investigations null and void.

Typically, in conducting forensic investigations, or incident response, a number of factors could hamper the admissibility of the results or evidence. In this paper, we have categorized the issues into three (3) externalities, which are:

- Intangibility – issues involving the RAM and memory analysis;
- Complications – issues involving anti-forensics, which can divert the focus of the investigations, or hide key evidential artefacts using such techniques as Steganography, Attention-Deficit-Disorder (ADD), Dementia, etc.;
- Inconsistencies – issues involving procedures, usage of tools and techniques, imaging of drives, chain-of-custody, the Locard Exchange Principles, etc.

The Locard Exchange Principle stipulates that the perpetrator is likely to leave traces and/or carry some evidentiary artefacts at the crime scene [4]. In expantiating the Locard's digital forensic cyber exchange principle, [5] posited that the traceable artefacts upon the cyber-incident, requires delving deeper into the compromised computing resource to adduce the evidence. This activity is challenging and by itself could introduce further complications and/or inconsistencies, if strict best practices are not adhered to.

2 Problem Formulation

Cyber-crime issues have assumed global dimension, with the perpetrators' profile ever becoming sophisticated [6], whilst forensic examiners and law enforcers are saddled with challenges in jurisdiction and prosecutorial difficulties [7]. Whereas emerging technologies are providing new opportunities for cyber criminals, new challenges and concerns for forensic examiners and law enforcement officers are emerging [8, 9]. The extent of cyber-crime has increased in recent times, and experts believe if nothing done to curb the menace, its impact is likely to be catastrophic in future [2].

Reported and/or documented cases of cyber-crimes and incidents are overwhelming and the quantum of dollar losses is gargantuan and mind-boggling. Interestingly, the reported cases during the period, 1991–1995, to the US CERT increased by almost 500% and the world-wide incidences had increased by over 700% [8]. Recently, the Malaysian CERT reported of 40.9% in fraud, 7.9% in malware, 5.3% in harassment, respective increases from 2013 to 2014 [10].

In a simulated penetration testing of US government computers, it was reported that about 65% successfully attacks occurred, with only 4% detected by the administrators [8]. It must be noted that lots of cyber incidents are not reported, and that unreported cases could also impede cyber-crime investigations.

In carrying out the incident response, the Forensic Examiner needs to answer a number of pertinent but daunting questions. For instance, "where did he obtain his/her tools from?"; "what standards and control measures or precautions are adhered to?"; "are the processes and procedures in conformity with laid down principles?"; "are there conformity or uniformity with his definitions, chain-of-custody and analysis – in respect of the admissibility of the digital evidence?"

The object of this study is to among others, carry out malware analysis, in order to determine the malware activities or operations, to comprehend the malware behavior, and analyze the workings of the malware codes. This study employs dynamic malware analysis, which can analyze activities in respect of threats attacking information assets in various states. For instance, attacks in transit or during transmission in networks and hosts, in storage or in file systems, in usage or in memory are all analyzed [11].

The paper is organized as follows: this introductory section deals with the background and problem formulation; the succeeding section deals with the literature review on digital forensic investigations and the externalities;

the following section deals with experiment and its approach; the following section presents the results and analysis. The implications of the findings are discussed and concluded.

3 Literature Review

This section reviews literature on digital forensic investigations and its challenges encountered, especially those that are likely to affect adversely the admissibility of the adduced evidence in court. Various concerns are at play here: from procedural through perceptions and technicalities.

Some of the issues raised to discredit digital forensic investigations include the lack of trained forensic analyst, proficiency testing, certifications, best practices, policies and procedures, laboratory standards, and accreditation, etc. [12].

3.1 Digital Forensics Investigation

Digital Forensic Investigation is basically incident response assessment to present digital evidence, which must be admissible in court, which may be used for either criminal or corporate and civil proceedings. The digital evidence artefacts often examined include, but not limited to, personal computers (PCs), laptops, notebooks, smartphones, cellphones, tablets, servers, GPS devices, Gaming Consoles, storage media, network devices and infrastructure, etc.

Some of the issues or cases usually being investigated include [13]:

- Use of smartphones to snap a picture or record a video, that was subsequently uploaded or shared unto a social network platform, e.g. Facebook, or sent via an email as an attachment;
- Use of computing resources to download an illicit image or file, e.g. child pornography, confidential corporate documents, etc.;
- Use of computing resources to commit financial crimes, fraud, money laundering, employee misconduct, copyright violations, distribution of inappropriate materials, etc. [14].

Another challenge with digital forensics is the perception of some law enforcement officers that digital forensics is merely an investigative tool, rather than a scientific evaluation [13]. This is more evident when due to organizational structural issues in the security agencies, officers are re-deployed periodically, without recourse to continuous experience leveraging in digital forensic investigations.

3.2 Intangibilities

Typically, intangibility is used to describe the ability to assess the value gained from engaging in an activity using any tangible evidence, e.g. software intangibility [15]. To what extent can the intangibility of cyber-crime complicate investigations and subsequently prosecution? Some difficulties encountered during investigations, especially in collecting evidence, can be associated with characteristic service attributes, which are either "intangibility, inseparability, perishability or heterogeneity" [16, p. 2].

In this study, we reckon malware activities as a computing service, albeit with negative network effects. Intangible nature of service(s) is such that it is performed and not easily measured. In describing the intangibility of services in computing, [16] related with the customer's change in his/her experiences, be it visible or not with change in computing resource state of operations. This introduces the potential for digital alteration in the evidence.

Evidentiary items are typically in both analog and digital formats [17], tangible and intangible forms, etc.

3.3 Complications

Generally, complications are difficulties or challenges emanating as a result of a certain circumstance or occurrence. In the context of digital forensic investigations, complications may arise from a cyber incident and the associated incident response. It may be legal, technical or ethical in nature. These complications can also emanate from the examiner or investigator, as well as the perpetrator, as he uses state-of-the-art techniques to circumvent the investigation.

The investigation becomes complicated even when one or a part of the compromised systems are geo-located in another jurisdiction [18]. Complications like this could sometimes end the investigations abruptly, though recent cases have seen the collaboration and cooperation of international law enforcement agencies, particularly the INTERPOL [19].

Another complication may arise with regards to the requirement of the forensic examiner to possess a state-accredited license in order to have the evidence admissible [20].

Some complications in investigating incident response or cyber-crimes arise out of objects or artefacts hiding using anti-forensic toolkits. Especially, in Windows based systems, since the Memory Analyzer tool has to run on the compromised machine, there's a potential for perpetrators to hide in memory dumps.

The use of anti-forensics are meant to thwart the investigations and to pollute the memory with fake artefacts. This study utilizes tools such as the Dementia (a DOS based tool for hiding "target artefacts in memory dumps – hides processes, process threads and [associated] connections" [21, p. 30]; Steganography – tools that facilitate hidden data in a carrier file or data; Attention Deficit Disorder (ADD) [22].

Microsoft Computer Online Forensic Evidence Extractor (COFEE) includes tools for password decryption, Internet history recovery and other data extraction. It also recovers data stored in volatile memory which could be lost if the computer were shut down.

3.4 Inconsistencies

Inconsistencies are anomalies, omissions, exaggerations or other loopholes found in forensic investigations contradictory to the evidence and which may compromise the value of the evidence or renders it inadmissible [20, 23].

These inconsistencies come in different forms. There are those associated with the compromised systems possibly due in part to the normal operations of the system (even in an uncompromised state) and that resulting from the compromises. These types of inconsistencies are described by [24] as temporal inconsistencies. They proposed a technique for detecting inconsistencies in digital forensics investigation.

Another type, jurisprudential inconsistencies may be due mainly to the lack of experiences of the judges on computer related crimes, cyber-crimes, or crimes committed with high-tech devices and/or against them, etc.

4 Methodology

In this section, we define the experimental set-up, its configurations, and the procedures and tools adopted for the exercise. Malware exploits utilized are clearly defined with its sources.

4.1 Experimental Set-up Laboratory

We installed a virtual environment using VirtualBox 5.0.4 for Windows hosts, from Oracle. This was for convenience and ease of use. Then, we installed Windows XP, both SP2 and SP3 packs. This was followed by a Linux operating system based toolkit, REMnux, for its robustness in respect of tools used for the analysis. In configuring the REMnux and the Windows servers, the default network setting is NAT (network address translation), but we configured it for "Host Only Adapter" and enabled promiscuous mode to "allow ALL".

It must be noted that, the set-up was carried out with a virtual-ware, so that the experiment does not interfere with the normal nor escalate malware to production environment.

The captured images and data were also analyzed using other tools, such as TriageIR (for incident response), LiveResponse, Volatility, etc. For the host analysis, we used tools such as the Magnet Forensics RAMCapture, FTKImager, MFT Dumper (for the Master File Table), AnalyzeMFT (to create the timelines), etc. For the network analysis, we used tools such as the WireShark, NetworkMiner, TCPDump, etc. For the memory analysis, we used mainly the Volatility toolkit.

REMnux is a toolkit, developed by Lenny Zeltser [11], to help researchers and computer forensic investigators to extract artefacts for analysis, with the view to produce credible (court admissible) digital evidence. REMnux contains a number of tools and commands for analyzing malicious softwares or malwares. REMnux is pre-configured to facilitate the investigation by prompting the examiner to select the data to be exported, and it is usually saved and stored in the same drive space from which the tool was launched.

We configured the REMnux to run a pseudo DNS service called Fakedns, a proxy (a NAT server) with SSL protocol and as a gateway for the Windows machines. We reconfigured the IPTables by writing a script that points DNS requests to the REMnux gateway. The Fakedns (or pseudo DNS server) runs on the REMnux and uses "whereis fakedns" Linux command.

−*sudo fakedns* 192.168.56.103 - running

*// started a web server using Apache; to as certain, we opened a web browser for confirmation;

*// in setting up the IPTables

- *sudo sysctl* − *w net.ipv4.ip _ forward*

−*sudo iptables* −*t nat* − *A prerouting* -1 *eth0* -*p tcp* - - *rdport*

With this setup, any DNS requests from the victim machine are re-directed to the fakedns server. The IPTables have also been re-configured as follows: TCP #80 and/or TCP #443 meant for HTTP and HTTPs services are re-directed to the pre-configured port TCP #8080.

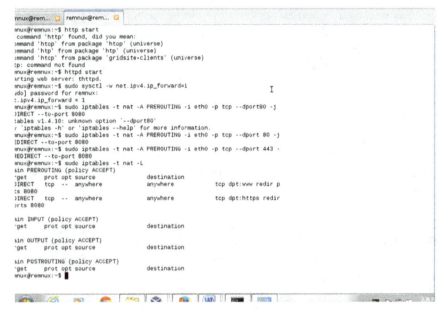

Figure 1   Configuration of the REMnux.

4.2 Malware Samples & Tools Used

Malware exploits or samples used were mainly taken from [11]; the Trojan .Stabuniq was specifically used in the experiment. The Trojan.Stabuniq was discovered by Symantec on December 17, 2012 [25]. It is a Trojan horse that steals confidential information from compromised systems. It is usually found in proxy and gateway servers. Upon infection, the Trojan masquerades amongst the existing files. It could affect a banking machines as well as home PCs. The typical attack vector used by the Trojan.Stabuniq is through emails and malicious websites, using phishing attacks [26].

FTKImager was used to capture disk image for analysis; RAMCapture tool was used to capture a memory image for analysis; and WireShark and TCPDump, running on the REMnux, were used to capture network packets.

Amongst the tools utilized in the REMnux suite is a process explorer and monitor. These are advanced monitoring tools for Windows, which monitors in real-time system activities, including Registry and process threads and activities [27].

The monitoring capabilities include filtering and furnishing of a compre-hensive list of event parameters, such as "session IDs, full thread stacks"

[27, p. 1] as well as logs and associated timestamps for boots and processes. We also used SysInternals utilities such as Regshot to facilitate the snap shots of processes running before and after the malware is launched. This helps in the comparison and analysis of the differences that have occurred.

5 Results & Analysis

This section presents the results of the experiment and some analytics of the malware behavior. First, we used the Magnet Foresnic RAMCapture tool to capture the memory image for analysis with the Volatility toolkit. The RAM image profile of the WIN SP2x86, image date and time created as 2015–10–11 13:47:40 UTC.

By inputting the Linux command: *remnux@remnux*: –$ ifconfig depicted the IP address of the REMnux: *eth0*: *link encap*: *Ethernet Hwaddr inet addr*: 192.168.56.102, which serves as a gateway, an internal DNS, running fakeDNS and WireShark.

Figure 2 depicts the system status before the launching of the malware exploits, whereas Figure 3 shows the system upon infection. Figure 3 also shows the network connections for the malware. The malware first contacted

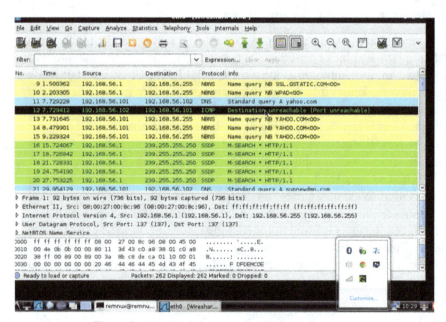

Figure 2   WireShark Report before Malware Infection.

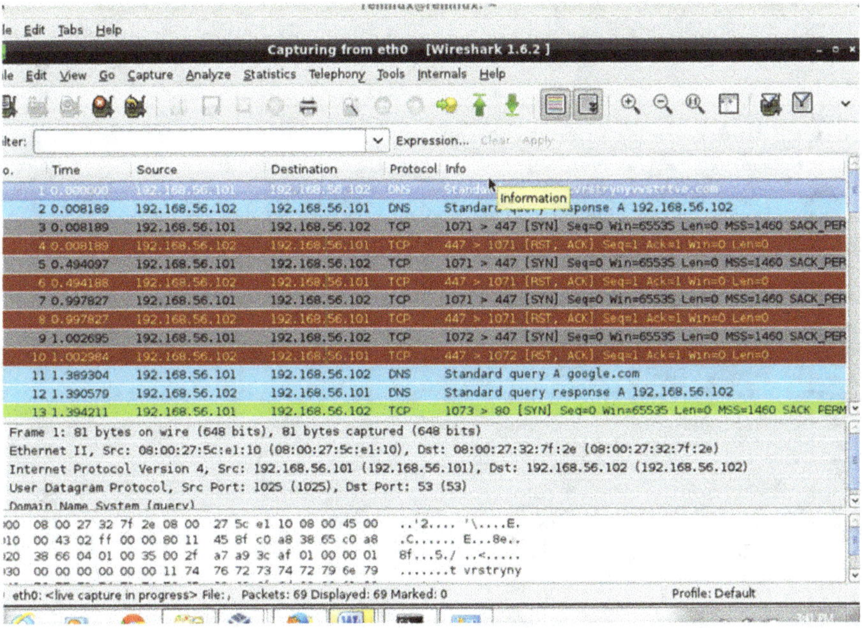

Figure 3   WireShark Report after Malware Infection.

the DNS server after an infection to a site www.tvrstrynyvwstrtve.com which is a suspected domain.Carrying out further malware analysis reveals that, the Trojan.Stabuniq commonly affect x86 or 32-bit machines and compatibles, and Windows GUI based systems, with file type WIN32.exe.

This malware uses the iEXPLORE.EXE, a legitimate file, to launch the viruses with its associated destructive payload.

With clean RAM image we found iEXPLORER with PIDs 164 and 152 and WSCNTFY 1054.

Using the Volatility toolkit, we executed "malfind" command and we found iEXPLORER with PIDs 1356, 1676 and 1684 as infected.

As indicated, the experiment was carried out in a controlled environment. This allows us to observe the creation of new files in the victim system, as it copies itself under different names and paths (c.f. Figures 4 & 5). The malware interacts with specific registry keys and injects the code into the iEXPLORE.EXE process, in order to ensure its persistence in the infected system, even after reboots.

We also observe some network activity in some domains as the malware interacts with the compromised system.

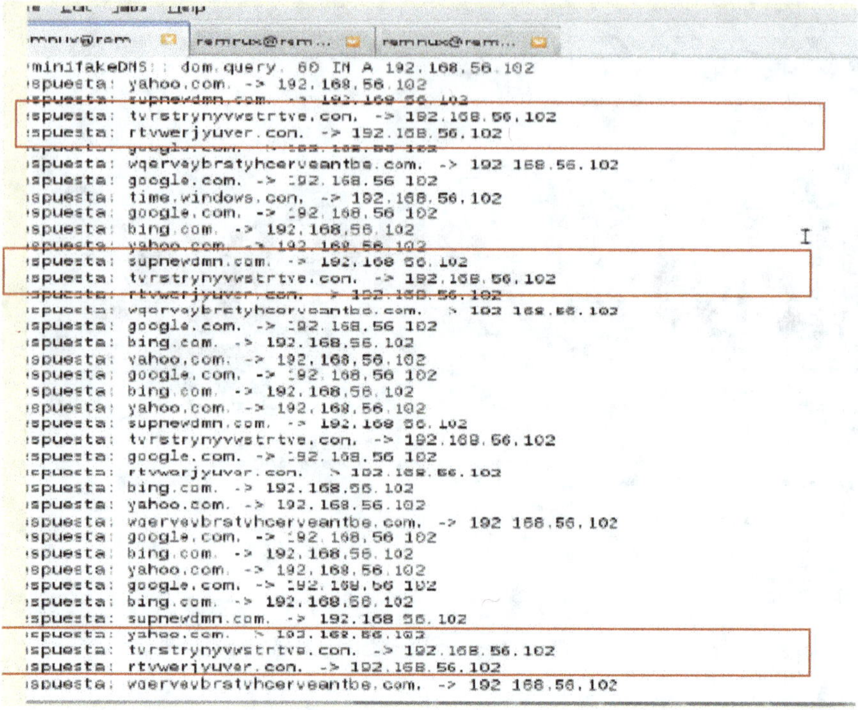

Figure 4   FakeDNS in action with Command & Control.

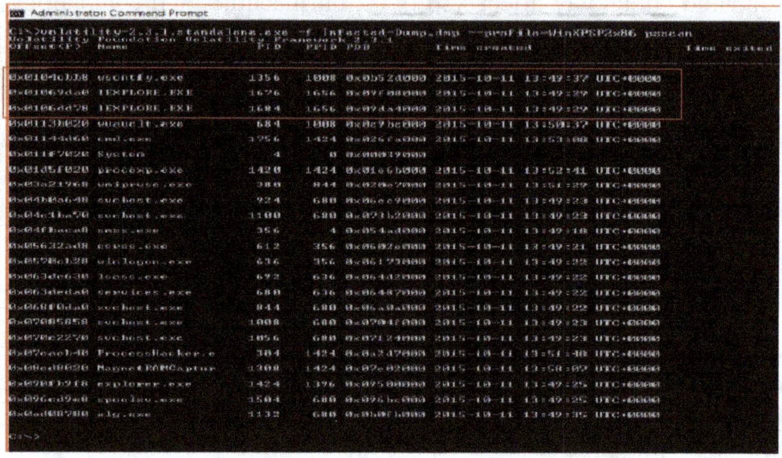

Figure 5   Volatility Plugin (PSSCAN).

Figure 4 depicts the fakeDNS in action to facilitate the malware operations with command and control privileges. Figure 5 on the other hand, depicts the results of a Volatility plugin called PSSCAN on the infected image that revealed the PIDs for iEXPLORER 1676 and 1684 with PPID 1656, WSCNTFY 1356 with a PPID 10008, which is SVCHOST.

The iEXPLORE.EXE is part of the Windows OS, for the Internet Explorer browser. Again, we herewith note that the browser was not active. However, the analysis shows the iEXPLORE.EXE running at the background with multiple iEXPLORE.EXE processes [28]. This was an indication of the possible presence of a malware; in this case the Trojan.Stabuniq. Typical symptoms of an infected system include slow computer, slow Internet connection, multiple browser crashes, re-directing pop-ups, etc. [29, 30].

6 Conclusion

Basically, in digital forensic analysis the best practice is to make an image (and hash it) before embarking of the analysis. The original copy is kept safe, whilst the duplicated "exact" copy is analyzed or worked on. This way in the event of any mis-analysis, another copy can be obtained.

In using the analytical tools, we endeavored to recover the remnants of the infected files or deleted files. It must be noted that, we began the examination by first copying the compromised disk, bit-by-bit so that bad sectors, unallocated spaces and deleted files could be examined. Since the compromised system under this experiment is a Windows OS, we then examined the image using Master File Table (MFT) attributes. These include the filename, time stamps (especially, the last modified, or accessed times), as well as the index entries for folders [18].

It was observed that some viruses don't work on the virtual environment, i.e. they recognize that possibly malware analysis will be carried out and so blocks the vmware. Some viruses upon launching appear and after a couple of seconds disappear.

We set off in this study to explore the extent of externalities, such as intangibility, complications and inconsistencies, impacting on cyber-crime investigations. We simulated an experimental cyber-crime in which a system had been adversely impacted with malicious codes. Carefully, best practices were adhered to ensure that the investigation and analysis did not complicate the delicate "intangible" evidence of imputing that a malware attack has taken place. We have also demonstrated that the perpetrator could hide or conceal his tracks with some techniques, often referred to as anti-forensics. In essence, any of these externalities could either hamper investigations or even render the entire investigations null and void, and the evidence inadmissible in court.

Obviously, more work ought to be demonstrated in showcasing other ways anti-forensics are used to hide and evade apprehension in many cyber-crime cases.

References

[1] E. O. Yeboah-Boateng, Cyber-Security Challenges with SMEs in Developing Economies: Issues of Confidentiality, Integrity & Availability (CIA), 1 ed., Copenhagen: Institut for Elektroniske Systemer, Aalborg University, 2013.

[2] B. Cashell, W. D. Jackson, M. Jickling and B. Webel, "The Economic Impact of Cyber Attacks," US Congressional Reserach Service, 2004.

[3] A. Karran, J. Haggerty, D. Lamb, M. Taylor and D. Llewellyn-Jones, "A Social Network Discovery Model for Digital Forensics Investigations," in *6th International Workshop on Digital Forensics & Incident Analysis (WDFIA 2011)*, 2011.

[4] Forensic Handbook, "Forensic Handbook," 12 August 2012. [Online]. Available: www.forensichandbook.com/locards-exchange-priniciple/. [Accessed 7 October 2015].

[5] K. Zatyko and J. Bay, "The Digital Forensic Cyber Exchange Principle," *Digital Forensic Investigator (DFI)*, 14 December 2011.

[6] E. O. Yeboah-Boateng and P. M. Amanor, "Phishing, SMiShing & Vishing: An Assessment of Threats against Mobile Devices," *Journal of Emerging Trends in Computing and Information Sciences*, vol. 5, no. 4, pp. 297–307, April 2014.

[7] FBI IC3, "2014 Internet Crime Report," Federal Bureau of Investigations, Internet Crime Complaint Cneter (IC3), 2015.

[8] S. Charney and K. Alexander, "Computer Crime," Computer Crime Research Center (CCRC), 2002.

[9] PITAC, "Cyber-Security: A Crisis of Prioritization," National Coordination Office for Information Technology Research & Development, 2005.

[10] MyCERT, "MyCERT Quarterly Incident Statistics Summary Report," 2014.

[11] L. Zeltser, "Malware Sample Sources for Researchers," 2013. [Online]. Available: www.zeltser.com/malware-sample-sources/. [Accessed 24 September 2015].

[12] J. Moulin, "Digital Forensic: The Impact of Inconsistent Standards, Certifications and Accreditation," 29015.

[13] SWGDE, Scientifc Working Group on Digital Forensics (SWGDE), 2014.

[14] E. O. Yeboah-Boateng and E. B. Boadi, "An Assessment of Corporate Security Policy Violations Using Live Forensics Analysis," *International Journal of Cyber-Security & Digital Forensics (IJCSDF),* vol. 4, no. 11, pp. 1–10, 2013.

[15] Essays-Lab, "Buy Custom Computer Forensic Essay," May 2015. [Online]. Available: www.essays-lab.com/free-samples/Research/computer-forensic.html. [Accessed 5 October 2015].

[16] A. Okunoye, "Increase in Computing Capacity and its Influence on Service Provision," in 37th *Hawaii International Conference on System Sciences – 2004,* 2004.

[17] D. J. Price, "The Analog and Digtal World," in *Handbook of Digital & Multimedia Forensic Evidence,* J. Barbara, Ed., Humana Press, 2008, pp. 1–10.

[18] S. Bui, M. Enyeart and J. Luong, "Issues in Computer Forensics," 2003.

[19] INTERPOL, "INTERPOL and Trend Micro to Collaborate Against Cybercrime," International Police, 24 June 2013. [Online]. Available: www.interpol.int/News-and-media/News/2013/PR076. [Accessed 7 October 2015].

[20] D. Shoemaker and W. A. Conklin, Cybersecurity: The Essential Body of Knowledge, Cengage Learning, Thomson Course Technology, 2011.

[21] L. Milkovic, "Defeating Windows Memory Forensics (29c3)," INFIGO, 2012.

[22] J. Stuttgen and M. Cohen, "Anti-Forensic Resilient Memory Acquisition," *Digital Investigation,* vol. 10, pp. 105–115, 2013.

[23] B. Nelson, A. Phillips, F. Enfinger and C. Steuart, Guide to Computer Forensics and Investigations, Cengage Learning, Thomson Course Technology, 2004.

[24] A. Marrington, G. Mohay, A. Clark and H. Morarji, "Dealing with Temporal Inconsitency in Automated Computer Forensic Profiling," Information Security Institute, Queensland University of Technology, 2009.

[25] E. D. Lucia, "Stabuniq in Depth," 24 December 2012. [Online]. Available: www.contagiodump.blogspot.com/2012/12/dec/dec-2012-trojanstabuniq-samples.html. [Accessed 2 October 2015].

[26] C. Robertson, "Indicators of Compromise in Memory Forensics," SANS Institute InfoSec Reading Room, 2013.

[27] M. Russinovich, "Process Monitor v3.2.," TechNet, 26 May 2015. [Online]. Available: www.technet.microsoft.com/en-us/library/bb896645.aspx. [Accessed 11 October 2015].

[28] M. Sirorski and A. Honig, Practical Malware Analysis: The Hands-on Guide to Dissecting Malicious Software, No Starch Press, 2012.

[29] Microsoft, "Malware Removal Guides: How to Remove Malware from Your Windows PC," Microsoft Corporation, 2014. [Online]. Available: www.malwareremovalguides.info/iexplorer-exe-is-running-in-background/. [Accessed 2 October 2015].

[30] Y.-M. Wang, R. Roussev, C. Verbowski, A. Johnson and D. Ladd, "AskStrider: What has Changed in My Machine Lately?," Microsoft Research, Microsoft Corporation, 2004.

[31] E. Casey, Handbook of Computer Crime Investigations: Forensic Tools and Technology, Academic Press, 2003.

[32] S. Chandra and R. K. Yadav, "Network Monitoring and Forensics," *International Journal of Computer Science and Mobile Computing,* vol. 2, no. 8, pp. 181–185, 2013.

[33] L. Volonino and I. Redpath, e-Discovery for Dummies, Wiley Publishing, Inc., 2010.

Biographies

E. O. Yeboah-Boateng is a senior lecturer and the Head (acting Dean), Faculty of Informatics, at the Ghana Technology University College (GTUC), in Accra. Ezer is an ICT Specialist and a Telecoms Engineer, an executive with over 25 years of corporate experience and about 9 years in academia. He has over 10 peer-reviewed international journal papers to his credit, and well cited in Google Scholar. His research focuses on cyber-security vulnerabilities, digital forensics investigations (DFI), cyber-crime and crimeware, cloud computing, Big data and fuzzy systems.

E. Akwa-Bonsu is a Cyber Security Expert and Researcher. Elvis is the Head of Intelligence at Detectware, a private cyber-security firm in Accra, Ghana. With 18 years of corporate experience, Elvis focuses on offensive, destructive, and defensive technology that affect and protect enterprises. He frequently speaks on the subject of security standards, penetration testing/auditing, digital investigations, attack techniques, wireless security, covert channel communications, network security monitoring, Packet Analysis, Malware Analysis, steganography, incident response, malware analysis, Honeypots, vulnerability analysis, virtualization, cloud computing security, business continuity and security awareness.

Factors Influencing the Continuance Use of Mobile Social Media: The Effect of Privacy Concerns

Kwame Simpe Ofori[1,*], Otu Larbi-Siaw[1,2], Eli Fianu[2],
Richard Eddie Gladjah[3] and Ezer Osei Yeboah Boateng[2]

[1]SMC University, Switzerland
[2]Ghana Technology University College, Ghana
[3]Ho Polytechnic, Ghana
*Corresponding Author: kwamesimpe@gmail.com

Received 31 August 2015; Accepted 20 November 2015;
Publication 22 January 2016

Abstract

With over 800 million active Whatsapp users, Mobile Social Networks (MSNs) have become one of the most vital means of social interactions, such as forming relationships and sharing information, sharing personal experiences. The mass adoption of MSN raises concerns about privacy and the risk of losing one's personal information due to personal data shared online. This paper sought to examine the role of Privacy Concerns in the continuance use of Mobile Social Media. The Effects of factors such as Perceived Ease of Use, Perceived Usefulness and Perceived Risk and Perceived Enjoyments on Satisfaction and Continuance intention were also explored. Survey data was collected from 262 students in Ghana Technology University College and analysed using the Partial Least Square approach to Structural Equation Modelling with the use of SmartPLS software. Results from the analysis showed that Perceived Usefulness, Perceived Risk and Perceived Enjoyment were significant predictors of Satisfaction. Satisfaction in turn was found to be a significant predictor of Continuance Intention. Satisfaction also mediated the paths between Perceived Risk, Privacy Concern and Continuance Intention. The results are discussed and practical implications drawn.

Journal of Cyber Security, Vol. 4, 193–212.
doi: 10.13052/jcsm2245-1439.426
© 2016 *River Publishers. All rights reserved.*

Keywords: Mobile Social Media Network, Privacy Concerns, Perceived Risk, Continuance Use.

1 Introduction

Mobile Social Media is fast becoming the number one medium for social interactions. This development is fuelled by the proliferation of mobile devices and the ever increasing 3G and 4G penetration. The mass adoption of this service presents some challenges with regards to privacy and the risk of losing one's personal information. Most Mobile Social Media Applications collect personal information such as user's demographics, preferences, and location from users. These information in the hands of wrong person could result in identity theft and other illegal use of such information. Privacy concerns could be a major reason why people do not adopt or discontinue after initial adoption.

Retaining users and facilitating their continuance use is essential to the success of any information system (Zhou & Li, 2014). This is because the cost of acquiring new customers is about five times that of retaining old ones (Bhattacherjee, 2001; Spiller, Vlasic, & Yetton, 2007; Vatanasombut, Igbaria, Stylianou, & Rodgers, 2008). For service providers to be able to recoup the investment made installing and launching the service it is therefore important that users of the service do not discontinue after initial adoption. The mobile social media market is quite competitive as they offer similar services and features (Zhou & Li, 2014). It is also quite easy for users to switch from one MSN to the other since a lot of the applications can be downloaded for free. The continuance usage by user therefore can be a source of competitive advantage for these service providers. Social media networks like hi5,MiSpace and Blackplanet which were unable to retain their users have lost out in this competitive arena. Despite the need to understand users' continuance intention to Mobile Social Network there is no research on this issue in the Ghanaian context.

This endeavoured to study fills this knowledge gap by seeking to identify the factors affecting users' intention to continue using Mobile Social Networks, and to analyse the relationships between these factors. We incorporated perceived enjoyment, perceived risk and privacy concern into the Expectancy Confirmation Theory of IS continuance. Results of this study extends the literature on continuance in the Mobile Social Network environment.

In the next section we presents a review of previous literature related to continuance usage of mobile social network and our proposed research model. In Section 3, the research methodology used for the study is described. The results of structural equation modelling (SEM) are presented in Section 4. Section 5 discusses the results of our study and highlights some theoretical and practical implications of our findings. We also identify some limitations of this study and provide directions for future research.

2 Theoretical Background/Concepts and Hypotheses Development

2.1 Theory of Reasoned Action (TRA)

The theory of reasoned action is one of the earliest theories applied to information systems adoption and has enjoyed prominence in academic literature. The theory was originally developed in 1967 but popularised by Ajzen & Fishbein, (1980). According to the theory a person's behaviour has two motivational components. First one's attitude towards performing the behaviour and their concern about what other people who matter to them would think about that behaviour (Social Norm). Ajzen & Fishbein (1980) observed that a person's intention to perform a behaviour was the only antecedent of the actual performance of the said behaviour. They also noted that behavioural intention was jointly determined by attitude and social norms.

Attitude represents a person's positive or negative feelings toward performing the said behaviour and is formed after an assessment of the consequences and the impact of the consequences of performing the said behaviour. Subjective norm on the other hand is a person's perception that people who are important to him/her think he/she should or should not perform that action (Ajzen & Fishbein, 1980). A pictorial view of the model is presented in Figure 1.

$$\mathbf{BI = A + SN} \tag{1}$$

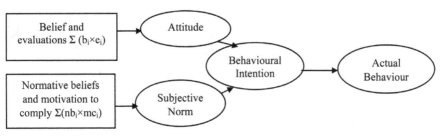

Figure 1 Theory of reasoned action (TRA) (Ajzen and Fishbein, 1980).

2.2 Technology Acceptance Model (TAM)

The Technology Acceptance Model (TAM) proposed by Davis (1989) was deduced from the Theory of Reasoned Action (TRA). Though TRA remains a general theory to clarify general human conduct, TAM is particularly suited for information system usage. TAM was initially created to comprehend the causal connection between outer variables and client acknowledgment of PC-based applications. TAM has been generally utilized as a hypothetical system as a part of the late studies to clarify technology acceptance (Moon and Kim, 2001; Gillenson and Sherrell, 2002; Koufaris, 2002; McCloskey, 2004; Chen).

As exhibited in Figure 2, the model sets that genuine utilization is controlled by user behavioural intention to use (BIU), which is impacted by their attitude (A) and the conviction of perceived usefulness (PU). Users attitude, which reflects positive or unfavourable emotions towards utilizing the IS framework, is resolved mutually by perceived usefulness (PU) and perceived ease of use (PEOU). PU, thus, is impacted by PEOU and external variables. The external variables may incorporate system design elements, preparing, documentation and client support, and so on. The rationale innate in the TAM is that the easier mastery of the technology, the more helpful it is seen to be, in this way prompting more inspirational demeanour and more positive attitude and greater intention towards utilizing the technology and thus ensuring more prominent use of the technology.

2.3 Expectation Confirmation Model of Information Systems Continuance (ECM-IS)

Bhattacherjee (2001) extended the work on Expectation (Dis)confirmation Theory (ECT) (Oliver & Burke, 1999; Oliver, 1999; Oliver & Linda, 1981; Oliver, 1980) to study information systems, specifically online banking consumers continuance behave. He likened IS continuance decision in the

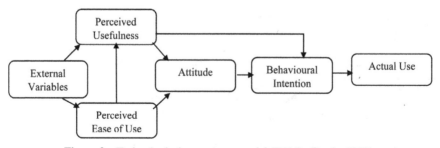

Figure 2   Technological acceptance model (TAM), (Davis, 1989).

ECM-IS to repurchase decisions of consumers in the ECT because both decisions follow the sequences of (1) making initial adoption/acceptance or purchase decision, (2) initially using the product/service and (3) making ex-post decision to continue or reverse the initial decision. The model proposes that a user's satisfaction derived from the user of an IS product/service leads to its continuance use. Satisfaction is the ex-post evaluation of the user's initial experience represented by a positive feeling, indifference or negative feeling. According to Bhattacherjee (2001) a user's satisfaction with information system (IS) is dependent on the extent to which their initial expectations of the performance of the IS is confirmed (disconfirmed) during actual use. He proposed that satisfaction is predicted by perceived usefulness and confirmation. Perceived usefulness is the user's perception of the benefits they expect to derive from using the system and confirmation is the congruence between how the user expects the system to perform and how it actually performs. (Bhattacherjee, 2001).

2.4 Privacy Concern

Privacy concerns reflects a user's perception of how their personal information is being used. Social media network service providers collect information concerning users. They also hold information on the users' interactions with other users. Users of social networks may be concerned about how providers may collect store and use their personal information. If users feel that this information may leaked or that they can be tracked due to a privacy violation then, they may be reluctant to use the system. Users with high privacy concerns are of the opinion that giving up their personal information may result in some privacy risk to them. It may require greater levels of trust in the system to have these users adopt the system. Privacy concern has been found to directly affect behavioural intentions of users in different contexts (Akhter, 2014; Arpaci, Kilicer, & Bardakci, 2015; Bansal, Zahedi, & Gefen, 2010; Gao, Waechter, & Bai, 2015; Li, 2014; McCole, Ramsey, & Williams, 2010; Zhou & Li, 2014; Zhou, 2011a). Aside the direct effect that privacy concern has on user behaviour, it has be shown to have indirect effects on user behaviour through the mediating roles of trust, perceived risk and perceived usefulness (Gashami, Chang, Rho, & Park, 2015).

2.5 Hypotheses Development

Perceived Ease of Use (PEOU): The degree to which a person believes that using a particular system would be free from effort (Davis, 1989). Davis

(1989) argued that the usefulness of a technology is dependent on how easy it is to use. In other words the easier it is for a user to interact with mobile social networks the more useful users would perceive it to be. The causal link between perceived ease of use and perceived usefulness has been tested and validated by previous researchers (Jongchul & Sung-Joon, 2014; Park, Rhoads, Hou, & Lee, 2014; Terzis, Moridis, & Economides, 2013). We therefore posit that:

H_{1a}: Perceived Ease of Use positively affects Perceived Usefulness.
H_{1b}: Perceived Ease of Use positively affects Satisfaction

Perceived Usefulness (PU): The salient belief that the degree to which a person uses a particular system would enhance his or her job performance (Davis, 1989). According to Burke (1997), perceived usefulness is the primary prerequisite for mass market technology acceptance, which solely depends on consumers' expectations about how technology can improve and simplify their lives. The expectation confirmation model of IS continuance also posits that when the system performs to the expectations of users they tend to get satisfied and in turn continue to use the IS (Bhattacherjee, 2001). Mobile social networks enable users to interact and share experiences with others anywhere and at any time. Users whose expectations about the performance of mobile social media are met are expected to be satisfied leading to their continuance use. Based on the above arguments we propose the following hypothesis:

H_2: Perceived Usefulness positively affects Satisfaction

Perceived Enjoyment (ENJ): Zhou (2011b) and Moon & Kim (2001) view perceived enjoyment as an intrinsic source of motivation, referring to the degree of pleasure derived from using the system. When users adopt mobile social networks they expect to enjoy the service. As these expectations are met they tend to get satisfied with the service (Zhou, 2011b). The effect of perceived enjoyment on user behaviour has been studied extensively and has been supported (Cheng, 2012; Jongchul & Sung-Joon, 2014; Koufaris, 2002; Zhou, 2011b). In line with the above literature we propose the following hypothesis:

H_3: Perceived Enjoyment positively affects Satisfaction

Perceived Risk (PR): The spatio-temporal nature of mobile applications and the unpredictability of the Internet infrastructure give rise to an implicit uncertainty with internet based systems (Pavlou, 2003). Various forms of risk exist in internet systems such as time risks, performance risk, privacy risk, financial risk, etc. In mobile social network, users have to contend with the risk of losing personal information either voluntarily or involuntarily due to theft. In this research we are concerned with privacy risk such as the risk of losing personal information shared with friends and the uncertainty associated with disclosing personal information to the service provider. We expect that users with high perceptions of privacy risk will feel lack of control and therefore would not be satisfied with the service and would not want to continue using it. We go forward to propose the following hypotheses:

H_{4a}: Perceived Risk negatively affects Satisfaction

H_{4b}: Perceived Risk negatively affects Continuance Intention

Privacy Concerns (PRC): Privacy is the ability of an individual or group to seclude themselves or information about themselves and thereby reveal themselves selectively. The advent of the Web 2.0 has caused social profiling and is a growing concern for internet privacy. Concerns about ones' privacy refers to his/her concern that his/her privacy may be compromised (Zhou, 2011a). Privacy concern has been identified to have a direct significant effect on behavioural intention (Bansal et al., 2010; Gashami et al., 2015; Liu, Marchewka, Lu, & Yu, 2005). Privacy concern has also been seen to have indirect effects on behavioural intention through trust (Bansal et al., 2010; Gashami et al., 2015; Liu et al., 2005; Zhou, 2011a), perceived risk (Slyke, Shim, Johnson, & Jiang, 2006; Zhou & Li, 2014) and perceived usefulness (Kumar, Mohan, & Holowczak, 2008). In line with the expectancy theory users will disclose their personal information if they think the benefit associated with disclosure is greater than the risk (Culnan & Bies, 2003). From the above literature we expect that privacy concern will have an effect on perceived risk, satisfaction and continuance intention. Based on the above arguments we posit that:

H_{5a}: Privacy Concerns positively affect Perceived Risk

H_{5b}: Privacy Concerns negatively affects Satisfaction

H_{5c}: Privacy Concerns negatively affects Continuance Intention

Satisfaction: Satisfaction represents an amassed feeling developed with multiple interactions. As indicated by the expectation confirmation theory, satisfaction is a strong predictor of continuance intention (Bhattacherjee, 2001). The impact of satisfaction on user behaviour has been bolstered in various studies (Gao & Bai, 2014; Kuo, Wu, & Deng, 2009; Lin, 2012; Zhou, 2011b). We therefore posit that:

H_6: Satisfaction positively affects Continuance Intention

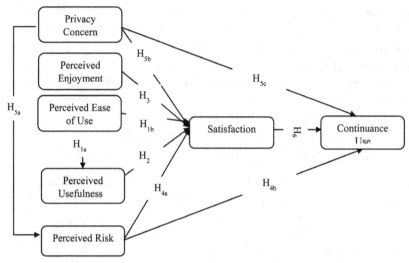

Figure 3   Research model.

3 Methodology

The measurement items for the latent variables used in this study were drawn from previous studies and the questions were reworded to fit the Mobile Social Network context. Perceived Usefulness and Perceived ease of use were from the Technology Acceptance Model, Satisfaction and Continuance from the ECM-IS model (Bhattacherjee, 2001), Perceived Enjoyment (Koufaris, 2002), Perceived Risk (Zhou, 2011a) and Privacy Concerns (Casaló, Flavián, & Guinalíu, 2007; Zhao, Koenig-Lewis, Hanmer-Lloyd, & Ward, 2010). Our measurement instrument had 23 items in all and items were presented in English and measured using a 5-point Likert scale anchored between 1 (Strongly Disagree) and 5 (Strongly Agree). Survey data was collected using paper-based questionnaires from 262 students in Ghana Technology University College and analysed using the PLS approach to Structural Equation Modelling with the use of SmartPLS software.

4 Results

4.1 Measurement Model

Results for the measurement model is presented in Tables 1 and 2. The measurement model is assessed based on reliability, convergent validity and discriminant validity. In evaluating indicator reliability the factor loadings of

Table 1 Factor loadings and cross loadings

		INT	PEOU	ENJ	PR	PU	PRC	SAT	CA	CR
CONTINUANCE INTENTION	INT1	**0.881**	0.202	0.246	-0.091	0.392	-0.138	0.571	**0.805**	**0.885**
	INT2	**0.793**	0.047	0.133	-0.157	0.277	-0.201	0.473		
	INT3	**0.869**	0.184	0.199	-0.028	0.332	-0.149	0.535		
PERCEIVED EASE OF USE	PEOU1	0.175	**0.839**	0.445	0.078	0.515	0.162	0.275	**0.748**	**0.855**
	PEOU2	0.059	**0.791**	0.369	0.014	0.416	0.055	0.203		
	PEOU3	0.183	**0.813**	0.364	0.016	0.428	0.085	0.241		
PERCEIVED ENJOYMENT	PENJ1	0.131	0.358	**0.822**	0.000	0.328	0.084	0.268	**0.872**	**0.912**
	PENJ2	0.212	0.419	**0.832**	0.025	0.440	0.160	0.262		
	PENJ3	0.253	0.441	**0.886**	-0.007	0.412	0.067	0.343		
	PENJ4	0.169	0.431	**0.856**	0.032	0.403	0.115	0.228		
PERCEIVED RISK	PR1	-0.036	0.121	0.091	**0.818**	0.043	0.151	-0.072	**0.832**	**0.870**
	PR2	-0.115	0.015	-0.016	**0.864**	-0.055	0.125	-0.164		
	PR3	-0.104	0.018	-0.015	**0.904**	-0.085	0.190	-0.169		
PERCEIVED USEFULNESS	PU1	0.349	0.513	0.455	-0.035	**0.903**	0.075	0.435	**0.848**	**0.908**
	PU2	0.340	0.472	0.432	-0.027	**0.857**	0.042	0.365		
	PU3	0.352	0.485	0.333	-0.071	**0.865**	-0.017	0.376		
PRIVACY CONCERNS	PRC1	-0.207	0.074	0.071	0.155	-0.001	**0.874**	-0.241	**0.87**	**0.91**
	PRC2	-0.131	0.166	0.133	0.127	0.062	**0.849**	-0.116		
	PRC3	-0.186	0.121	0.114	0.160	0.029	**0.822**	-0.140		
	PRC4	-0.094	0.098	0.108	0.172	0.062	**0.840**	-0.177		
SATISFACTION	SAT1	0.501	0.222	0.240	-0.160	0.326	-0.188	**0.855**	**0.799**	**0.882**
	SAT2	0.581	0.259	0.312	-0.168	0.378	-0.183	**0.862**		
	SAT3	0.490	0.270	0.282	-0.091	0.433	-0.155	**0.817**		

Table 2   Correlation matrix with square root of AVE

	INT	PEOU	ENJ	PR	PU	PRC	SAT	AVE
INT	0.848							0.720
PEOU	0.175	0.814						0.663
ENJ	0.230	0.486	0.849					0.722
PR	−0.105	0.048	0.012	0.863				0.744
PU	0.396	0.560	0.466	−0.050	0.876			0.767
PRC	−0.189	0.129	0.121	0.182	0.040	0.846		0.716
SAT	0.623	0.297	0.331	−0.166	0.449	−0.208	0.845	0.712

Note: Square roots of average variances extracted (AVEs) shown on diagonal

each indicator should be above 0.708 (Hair, Hult, Ringle, & Sarstedt, 2014). Indicator reliability was achieved since all factor loadings from Table 1 were above the threshold. Construct Reliability was also assessed using Cronbach's alpha and Composite Reliability. For constructs to be reliable Cronbach's alpha and Composite Reliability values should be greater than 0.7 (Hair et al., 2014; Nunnally & Bernstein, 1994). From Table 1 it is evident that all construct had Cronbach's alpha and Composite Reliabilities greater than 0.7, indicative of construct reliability. From Table 2 it can also be seen that the square root of the AVEs are greater than the cross correlations indicating that discriminant validity was satisfied.

Results from the analysis also showed that the factor loadings for each item is greater than the cross-loadings further providing support for discriminant validity. The Average Variance Extracted for each variable was above the threshold of 0.5 hence convergent validity was satisfied. Since the results from the analysis indicate that the constructs show sufficient levels of reliability, convergent validity and discriminant validity, a Structural Model was developed for further testing.

4.2 Structural Model Assessment

Results for structural model analysis are presented in Table 3. Perceived Ease of use was found to be a significant predictor of Perceived Usefulness ($\beta = 0.560$, P = 0.000) providing support for H_{1a}. Perceived Ease of Use was however found not to be directly affecting Satisfaction. Rather, it had an indirect effect on satisfaction through the mediating role of Perceived Usefulness. Perceived Usefulness ($\beta = 0.340$, P = 0.000), Perceived Enjoyment ($\beta = 0.174$, P = 0.009), Perceived Risk ($\beta = −0.112$, P = 0.078) and Privacy Concern ($\beta = −0.229$, P = 0.000) were also found to be significant predictors of Satisfaction providing support for H_2, H_3, H_{4a} and H_{5b} respectively.

Together they accounted for 29.3% of the variance in Satisfaction. Satisfaction ($\beta = 0.611$, $P = 0.000$) was also found to be a strong predictor of Continuance Intention. Privacy Concern and Perceived Risk were found not to directly affect Continuance Intention but indirectly through the

Table 3   Result for hypotheses testing

Hypotheses	Hypothesized Path	Path Coefficient	T Statistics	P Values	Result
H_{1a}	PEOU → PU	0.560	7.937	0.000	Supported
H_{1b}	PEOU → SAT	0.057	0.724	0.469	Not Supported
H_2	PU → SAT	0.340	3.854	0.000	Supported
H_3	ENJ → SAT	0.174	2.631	0.009	Supported
H_{4a}	PR → SAT	−0.112	1.768	0.078	Supported
H_{4b}	PR → INT	0.008	0.169	0.866	Not Supported
H_{5a}	PRC → PR	0.182	2.684	0.008	Supported
H_{5b}	PRC → SAT	−0.229	4.297	0.00	Supported
H_{5c}	PRC → INT	−0.064	1.185	0.237	Not Supported
H_6	SAT → INT	0.611	9.509	0.000	Supported

*Significant at = 0.1, **Significant at = 0.05, ***Significant at = 0.01

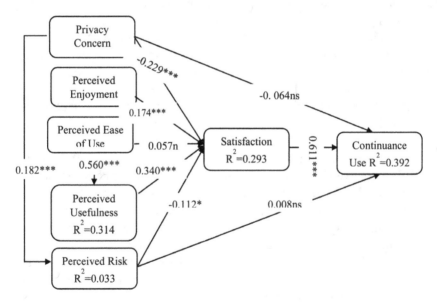

* Significant at = 0.1, ** Significant at = 0.05, ***Significant at = 0.01

Figure 4   PLS results for structural model.

mediating role of Satisfaction. About 40% of the variation in Continuance intention is accounted for by our model. Privacy Concern was also seen to have a significant effect on Perceived Risk.

Structural equation modelling was used to examine ten (10) hypotheses proposed for the study and as indicated in Table 3, with the exception if H_{1b}, H_{4b} and H_{5c}, all the other hypotheses were supported. Properties of the causal paths, including standardized path coefficients and hypotheses testing results in the hypothesized model are presented in Table 3.

5 Discussions

Mobile Social Networks are gaining popularity among mobile users because of the ubiquitous presence it offers users. The speedy growth in the number of users currently using Mobile Social Network has necessitated research in the adoption and continuance intentions of customers.

In this research we sought to examine the factors that affect continuance intention of Mobile Social Network users by integrating constructs from the Expectancy Confirmation Model of IS continuance with perceived risk, privacy concerns, and perceived enjoyment. Our results provide insight into the relationship between continuance intention and its antecedents and offer two contributions. (1) User satisfaction with Mobile Social Networks is derived from perceived usefulness, perceived risk, perceived enjoyment and privacy concerns. (2) Users' continuance intentions is derived only from satisfaction, supporting the work of (Bhattacherjee, 2001).

In line with previous studies of our results indicate that Perceived Usefulness has significant influence on Satisfaction (Bhattacherjee, 2001; Kim, 2011). Perceived Ease of Use was also found to have significant impact on Perceived Usefulness providing support for (Lim, Al-Aali, Heinrichs, & Lim, 2013). Perceived Ease of Use had no significant impact on Satisfaction, this could probably be due to the fact that our sample consisted of young university students, most of whom are technology savvy. Perceived Enjoyment was also found to significantly predict Satisfaction corroborating previous studies (Jongchul & Sung-Joon, 2014; Zhou, 2011b). Mobile social networks have a variety of entertaining services such sharing videos and pictures. Users adopt mobile social media networks to interact with others and also to derive some enjoyment. From the results we can infer that as expectations of enjoyment is fulfilled they would become satisfied and continue to use the service.As expected we also found negative relationships between Perceived Risk, Privacy Concerns and Satisfaction. Implying that as users became aware of the privacy risks associated with social media network they were less likely to be satisfied.

Users provide a lot of personal information when they register on the social media network. Further, as they use the service, they post and share a lot of personal information which may be kept on the servers of the service providers. Some social networking apps can also collect location information from users to provide them location-based services. When users have the impression that these information may not be handled properly they may choose not to continue using the service. Service providers could post their privacy policy as this has been seen to engender trust in the service and reduce privacy concerns (Wu, Huang, Yen, & Popova, 2012).

Interestingly both perceived risk and privacy concern did not have any significant direct effect on continuance intention. However, through the mediating role of satisfaction both variables were seen to have an indirect effect on continuance intention. In line with previous studied we found a strong relationship between satisfaction and continuance intention (Bhattacherjee, 2001; Yoon & Roland, 2015; Yuan, Liu, Yao, & Liu, 2014).

5.1 Implications

From a theoretical standpoint, this study integrates the IS Continuance model with Privacy Concern, Perceived Enjoyment and Perceived Risk to explore the continuance use of mobile social networks. Previous works have concentrated on user adoption, but as noted earlier to ensure the sustenance of these social network provider, users must continue using the service. Also, extant research has focused on continuance use of social networks in the western world. However, not much has been done in the Ghanaian context. Our work therefore extends knowledge in continuance usage of IS services. Our results indicate that Privacy Concern has a significant negative effect on User Satisfaction. Perceive Usefulness, Perceive Enjoyment and Perceived Risk were also found to be significant in prediction Satisfaction.

From a practical perspective, the results imply that mobile social network service providers need to consider Privacy Concern, Perceived Enjoyment Perceived Risk and Perceived Usefulness if they want to make their users satisfied. This would ultimately lead to their continued use of the system. Also, since Privacy Concern was seen to have a strong negative influence on satisfaction service providers must adopt measures such as privacy policies and seals to lower the privacy concerns of users. They could also use reputation to engender trust and reduce privacy risk. Contrary to expectation and prior research Perceived Risk and Privacy Concerns did not have significant direct effects on Continuance Use but had indirect effect through satisfaction.

5.2 Limitation and Direction for Future Research

Even though the study found some interesting result that confirm previous studies a few limitations have to be considered when interpreting and generalizing results to other social media technologies and other countries. First the study collected data from university students and not the wider population. Although research has identified students as a representable part of the population, in the future researchers can consider using a different sample to re-test/validate the research model. Secondly the study did not consider the influence of demographic variables like age, gender, educational status, etc. In other studies these variables have been found to moderate relationship between factors that influence continuance. It would therefore be interesting to explore their effects. Finally our study used a cross-sectional design and since user behaviour is dynamic it would be exciting to consider a longitudinal design in future studies.

References

[1] Ajzen, I., & Fishbein, M. (1980). *Understanding attitudes and predicting social behavior. Englewood Cliffs NY Prentice Hall*

[2] Akhter, S. H. (2014). Privacy concern and online transactions: The impact of internet self-efficacy and internet involvement. *Journal of Consumer Marketing, 31*(2), 118–125. http://doi.org/10.1108/JCM-06-2013-0606

[3] Arpaci, I., Kilicer, K., & Bardakci, S. (2015). Effects of security and privacy concerns on educational use of cloud services. *Computers in Human Behavior, 45*, 93–98. http://doi.org/10.1016/j.chb.2014.11.075

[4] Bansal, G., Zahedi, F. "Mariam," & Gefen, D. (2010). The impact of personal dispositions on information sensitivity, privacy concern and trust in disclosing health information online. *Decision Support Systems, 49*(2), 138–150. http://doi.org/10.1016/j.dss.2010.01.010

[5] Bhattacherjee, A. (2001). Understanding information system continuance: An Expectation confirmation model. *MIS Quarterly*, 25(3), 351–370.

[6] Burke, R. R. (1997). Do You See What I See? The Future of Virtual Shopping. *Journal of the Academy of Marketing Science, 25*(4), 352–360. http://doi.org/10.1177/0092070397254007

[7] Casaló, L. V., Flavián, C., & Guinalíu, M. (2007). The role of security, privacy, usability and reputation in the development of online banking. *Online Information Review, 31*(5), 583–603. http://doi.org/10.1108/14684520710832315

[8] Cheng, Y. (2012). Effects of quality antecedents on e-learning acceptance. *Internet Research, 22*(3), 361–390. http://doi.org/10.1108/106622 41211235699

[9] Culnan, M. J., & Bies, R. J. (2003). Consumer Privacy: Balancing Economic and Justice Considerations. *Journal of Social Issues, 59*(2), 323–342. http://doi.org/10.1111/1540-4560.00067

[10] Davis, F. D. (1989). Perceived usefulness, perceived ease of use, and user acceptance of information technology. *MIS Quarterly*, 13(3), 319–340. http://doi.org/10.2307/249008

[11] Gao, L., & Bai, X. (2014). An empirical study on continuance intention of mobile social networking services. *Asia Pacific Journal of Marketing and Logistics, 26*(2), 168–189. http://doi.org/10.1108/APJML-07-2013-0086

[12] Gao, L., Waechter, K. A., & Bai, X. (2015). Understanding consumers' continuance intention towards mobile purchase: A theoretical framework and empirical study – A case of China. *Computers in Human Behavior, 53*, 249–262. http://doi.org/10.1016/j.chb.2015.07.014

[13] Gashami, J. P. G., Chang, Y., Rho, J. J., & Park, M.-C. (2015). Privacy concerns and benefits in SaaS adoption by individual users: A trade-off approach. *Information Development*, (5). http://doi.org/10.1177/02666 66915571428

[14] Hair, J. F. J., Hult, G. T. M., Ringle, C., & Sarstedt, M. (2014). *A Primer on Partial Least Squares Structural Equation Modeling (PLS-SEM)*. Thousand Oaks: Sage Publications. http://doi.org/10.1016/j.lrp.2013.01.002

[15] Jongchul, O., & Sung-Joon, Y. (2014). Validation of Haptic Enabling Technology Acceptance Model (HE-TAM): Integration of IDT and TAM. *Telematics and Informatics, 31*(4), 585–596. http://doi.org/10.101 6/j.tele.2014.01.002

[16] Kim, B. (2011). Understanding Antecedents of Continuance Intention in Social-Networking Services. *Cyberpsychology, Behavior, and Social Networking, 14*(4), 199–205. http://doi.org/10.1089/cyber.2010.0009

[17] Koufaris, M. (2002). Applying the Technology Acceptance Model and flow theory to online consumer behavior. *Journal of Information Systems Research, 13*(2), 205–223. http://doi.org/10.1287/isre.13.2.205.83

[18] Kumar, N., Mohan, K., & Holowczak, R. (2008). Locking the door but leaving the computer vulnerable: Factors inhibiting home users' adoption of software firewalls. *Decision Support Systems, 46*(1), 254–264. http://doi.org/10.1016/j.dss.2008.06.010

[19] Kuo, Y. F., Wu, C. M., & Deng, W. J. (2009). The relationships among service quality, perceived value, customer satisfaction, and post-purchase intention in mobile value-added services. *Computers in Human Behavior, 25*(4), 887–896. http://doi.org/10.1016/j.chb.2009.03.003

[20] Li, Y. (2014). The impact of disposition to privacy, website reputation and website familiarity on information privacy concerns. *Decision Support Systems, 57*(1), 343–354. http://doi.org/10.1016/j.dss.2013.09.018

[21] Lim, J.-S., Al-Aali, A., Heinrichs, J. H., & Lim, K.-S. (2013). Testing alternative models of individuals' social media involvement and satisfaction. *Computers in Human Behavior, 29*(6), 2816–2828. http://doi.org/10.1016/j.chb.2013.07.022

[22] Lin, W. S. (2012). Perceived fit and satisfaction on web learning performance: IS continuance intention and task-technology fit perspectives. *International Journal of Human Computer Studies, 70*(7), 498–507. http://doi.org/10.1016/j.ijhcs.2012.01.006

[23] Liu, C., Marchewka, J. T., Lu, J., & Yu, C.-S. (2005). Beyond concern—a privacy-trust-behavioral intention model of electronic commerce. *Information & Management, 42*(2), 289–304. http://doi.org/10.1016/j.im.2004.01.003

[24] McCole, P., Ramsey, E., & Williams, J. (2010). Trust considerations on attitudes towards online purchasing: The moderating effect of privacy and security concerns. *Journal of Business Research, 63*(9–10), 1018–1024. http://doi.org/10.1016/j.jbusres.2009.02.025

[25] Moon, J. W., & Kim, Y. G. (2001). Extending the TAM for a World-Wide-Web context. *Information and Management, 38*(4), 217–230. http://doi.org/10.1016/S0378-7206(00)00061-6

[26] Nunnally, J., & Bernstein, I. (1994). *Psychometric Theory* (3rd ed.). New York: McGraw-Hill.

[27] Oliver, R. L. (1980). A Cognitive Model of the Antecedents and Consequences of Satisfaction Decisions. *Journal of Marketing Research, 17*(4), 460–469.

[28] Oliver, R. L. (1999). Whence consumer loyalty? *Journal of Marketing.* Retrieved from http://search.ebscohost.com/login.aspx?direct=true& profile=ehost&scope=site&authtype=crawler&jrnl=00222429&AN=24 44274&h=gx9looLpWeEmSRwxdnVVTF0cZFXHLX%2FfSknEYzpS VlrFtQSvbNOKhqnVEe%2BLuhQgkQhS%2B7CGZ4l8YniBvZ%2Fl eg%3D%3D&crl=c

[29] Oliver, R. L., & Burke, R. R. (1999). Expectation Processes in Satisfaction Formation: A Field Study. *Journal of Service Research.* http://doi.org/10.1177/109467059913002

[30] Oliver, R. L., & Linda, G. (1981). Effect of satisfaction and its antecedents on consumer preference and intention. *Advances in Consumer Research, 8*(1), 88–93.

[31] Park, N., Rhoads, M., Hou, J., & Lee, K. M. (2014). Understanding the acceptance of teleconferencing systems among employees: An extension of the technology acceptance model. *Computers in Human Behavior, 39*, 118–127. http://doi.org/10.1016/j.chb.2014.05.048

[32] Pavlou, P. A. (2003). Consumer Acceptance of Electronic Commerce: Integrating Trust and Risk with the Technology Acceptance Model. *International Journal of Electronic Commerce, 7*(3), 34. http://doi.org/ 10.1080/10864415.2003.11044275

[33] Slyke, C. Van, Shim, J., Johnson, R., & Jiang, J. (2006). Concern for Information Privacy and Online consumer purchasing. *Journal of the Association for Information Systems, 7*(6), 415–444. Retrieved from http://ais.bepress.com/cgi/viewcontent.cgi?article=1266&context=jais

[34] Spiller, J., Vlasic, A., & Yetton, P. (2007). Post-adoption behavior of users of Internet Service Providers. *Information and Management, 44*(6), 513–523. http://doi.org/10.1016/j.im.2007.01.003

[35] Terzis, V., Moridis, C. N., & Economides, A. a. (2013). Continuance acceptance of computer based assessment through the integration of user's expectations and perceptions. *Computers and Education, 62*, 50–61. http://doi.org/10.1016/j.compedu.2012.10.018

[36] Vatanasombut, B., Igbaria, M., Stylianou, A. C., & Rodgers, W. (2008). Information systems continuance intention of web-based applications customers: The case of online banking. *Information and Management, 45*(7), 419–428. http://doi.org/10.1016/j.im.2008.03.005

[37] Wu, K. W., Huang, S. Y., Yen, D. C., & Popova, I. (2012). The effect of online privacy policy on consumer privacy concern and trust. *Computers in Human Behavior, 28*(3), 889–897. http://doi.org/10.1016/j.chb.2011. 12.008

[38] Yoon, C., & Roland, E. (2015). Understanding Continuance Use in Social Networking Services. *Journal of Computer Information Systems, 55*(2), 1–8.

[39] Yuan, S., Liu, Y., Yao, R., & Liu, J. (2014). An investigation of users' continuance intention towards mobile banking in China. *Information Development*. http://doi.org/10.1177/0266666914522140

[40] Zhao, A. L., Koenig-Lewis, N., Hanmer-Lloyd, S., & Ward, P. (2010). Adoption of internet banking services in China: Is it all about trust? *International Journal of Bank Marketing, 28*(1), 7–26. http://doi.org/10.1108/02652321011013562

[41] Zhou, T. (2011a). The impact of privacy concern on user adoption of location-based services. *Industrial Management & Data Systems, 111*(2), 212–226. http://doi.org/10.1108/02635571111115146

[42] Zhou, T. (2011b). Understanding mobile Internet continuance usage from the perspectives of UTAUT and flow. *Information Development, 27*(3), 207–218. http://doi.org/10.1177/0266666911414596

[43] Zhou, T., & Li, H. (2014). Understanding mobile SNS continuance usage in China from the perspectives of social influence and privacy concern. *Computers in Human Behavior, 37*, 283–289. http://doi.org/10.1016/j.chb.2014.05.008

Biographies

K. S. Ofori is a doctoral student at SMC University, Switzerland. He holds a Master of Science degree in Telecommunications Technology and a Postgraduate diploma in Business Administration. He is currently a lecturer in Information Security, Reliability Engineering and Data Communications. His research interests include PLS path modelling and technology adoption. Kwame can be reached at: kwamesimpe@gmail.com

E. Fianu is a faculty member in the Informatics Faculty of Ghana Technology University College (GTUC). His areas of lecturing are Database Systems, Systems Analysis and Design, Computer applications for Management, and E-Commerce. Prior to joining GTUC, he worked with Vodafone Ghana for 11 years as a Customer Experience, Sales and Service Provisioning Manager. He holds a Master of Science degree in Management Information Systems from Coventry University, and a Bachelor of Science Degree in Agricultural Economics from the University of Ghana. His research interests include Artificial Neural Networks, Technology Adoption, and e-Learning. Eli can be reached at: efianu@gmail.com

O. Larbi-Siaw is a rigorous problem solver, with a passion for designing systematic decision making systems. He is the Head of Department for Economics at Ghana Technology University College. His research interest is in the area of macroeconomics, innovation economics, game theory, artificial intelligence, digital analytics and finance. He is currently a PhD student at Swiss Management Center (SMC). Otu can be reached at: oLarbi-Siaw@gtuc.edu.gh

R. E. Gladjah is a lecturer in the Department of Mathematics and Statistics, Ho Polytechnic, Ghana. He had his first degree in Statistics and Economics from the University of Ghana and a Master of Science in Statistics from Regent University College of Science and Technology. He has seventeen years of teaching experience in Statistics and Computer Applications and specializes in Time Series, Multivariate and Categorical Data Analyses. His research interests include technology adoption and time series analysis. Eddie can be reached at: e1684@yahoo.com

E. O. Yeboah-Boateng is a senior lecturer and the Head (acting Dean), Faculty of Informatics, at the Ghana Technology University College (GTUC), in Accra. Ezer is a Telecoms Engineer and an ICT Specialist; an executive with over 20 years of corporate experience and about 8 years in academia. He has over 10 peer-reviewed international journal papers to his credit. His research focuses on cyber-security vulnerabilities, digital forensics, cyber-crime, cloud computing, Big data and fuzzy systems. He can be reached at: eyeboah-boateng@gtuc.edu.gh

Confidentiality in Online Social Networks; A Trust-based Approach

Vedashree K. Takalkar and Parikshit N. Mahalle

Smt. Kashibai Navale college of Engineering, Savitribai Phule Pune University, Pune, Maharashtra, India
Email: vedatakalkar@yahoo.com; aalborg.pnm@gmail.com

Received 2 November 2015; Accepted 4 December 2015;
Publication 22 January 2016

Abstract

Considering the growing popularity of the Online Social Networks, achieving data confidentiality from user's perspective has turned out to be a vital issue. A system using trust can provide access control for the data uploaded by the owner on the social network. The paper discusses various metrics to calculate the trust and evaluation of trust score to determine the trust an owner has with the friends in her social network. Also the paper proposes the architecture that will build this trust evaluation system. Hence, the data will be seen by the friends who are trusted and the motive to achieve data confidentiality is achieved using trust-based access control scheme. The paper also discusses the Trust Rule to achieve access control of the data. To the best of our knowledge, this is the first proposal that calculates trust based on experience, context information and interaction.

Keywords: Online Social Network, trust, trust score, access control, data confidentiality.

1 Introduction

Man is social animal. OSNs (Online Social Networks) are designed and developed for the people around the globe to interact with each other and get connected. This is a platform through which an OSN user develops his

Journal of Cyber Security, Vol. 4, 213–232.
doi: 10.13052/jcsm2245-1439.427
ⓒ 2016 *River Publishers. All rights reserved.*

own identity and interacts using this identity sharing his personal and public data with all the people connected to him called as friends. An OSN user gets connected with his friends, colleagues, friends-of-friends, relatives and even unknown people who then might become good friends. Thus OSNs were mainly developed for strengthening the already existing relations and establishing the new relations. To reap such benefits, people are using OSNs like Facebook, Twitter, Myspace, LinkedIn etc. Facebook statistics boasts to have [2] 3.17 billion active users. Thus, this figure explains the usage of the OSN. However, in the above scenario the data that is uploaded needs to be given a secured access. The survey of 325 Facebook [4] users claims that about 69.2% people keep their posts public and 7.7% people don't even know whether their posts are public or private. Also the survey [4] infers that only 19.4% people are concerned about the privacy policies that are used in Facebook. The conclusion from these statistics states that people are not much aware about the hazards and the problems that may be caused if the sensitive data is reached to the unintended users. Hence, some access control policies should be stated by the users for every data that they may upload. For example, if Alice uploads the photo of her family function, maybe she doesn't want it to be seen by the friends who she doesn't know much but still exist in her friend list. However, due to weak privacy concerns this motive is never fulfilled and the photo is accessed by the unintended friends. Also the present OSNs allows us either to keep the data public or private which are the two extreme cases that cannot provide the data confidentiality effectively [5]. Hence, giving access control of the data to the friends should be based upon some criteria. The paper discusses adding the flavour of trust to give access control mechanism. In real world all the relations are established on trust which mainly comes from the knowledge and experience of the person and his behaviour. The paper also discusses about the factors that are considered to calculate the trust among the owner of the profile and his friends. Depending upon the trust that the owner has towards his friends, the access of the data can be given. Thus, depending on this the selective display of data can be given to all the friends of the owner.

The Figure 1 shows the high level view of the OSN architecture. Here, the users shown are nothing but the user profiles that exists in the OSN to identify the real users. The data server is used to store the data uploaded by the users. The OSN provider [10] gives all the facilities that are needed by the OSN users. These services include functionalities like storage, maintenance and access of the data. The users interact with each other, upload data and communicate with each other. The architecture as shown in Figure 1 is a layered architecture of OSN where each layer shows the different tasks that are performed by OSN all

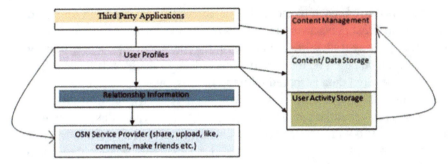

Figure 1 High level view of OSN.

together. On the top are the third party applications that are developed by the third party developers that are installed externally and manually by the users according to their own choice. Next layer is the user profiles which maintain the profiles of all the users. Relationship information is the social link that is shared between different users in the OSNs. It may include the various groups, friend, mutual friend and the other links. The last layer is the basic services that are provided by the OSN provider which are required by the users like sharing, uploading the data, commenting, following, liking, posting etc. These are all the OSN services and hence, act as the foundation for all the above layers. Along with this, all the layers are also dependent on the content manager and content storage. For, this all the data or the content needs to be managed and stored properly. Also, the user activity information is also needed to be stored like who liked which data, who commented on which data, who uploaded on which data etc. Hence, the user activity storage plays a vital role.

2 Motivation

It has been observed that users are not much aware of the privacy features available in OSN [2]. Hence, such ignorance leads to the data confidentiality attack and the secured data is accessed (shared, viewed) by the unintended people. Also, the people in order to increase their friend list tend to accept the friend request of strangers. Observing that the user is from same company and the action of declining the friend request should not hurt the person [11]; users tend to accept the friend request. Hence, all the friends in the friend list are not always the people who are best known or trusted. Hence, whenever any data like photos, status is uploaded owner may not want it to share with the people who exist in friend list but are actually the strangers. Hence, this may cause privacy breach to the highly secured data that has been uploaded by the owner. Trust plays a very important role in the daily life of man. All the transactions of day to day life are based on the real life trust. Hence, sharing of the data can be based on the trust the owner has for the friends in his friend list. Highly secured data is shared with the friends having greater trust.

Alice uploads a photo. She wants it to be seen by the friends who she trusts more as she feels that the data is more secured. It is obvious that the secured data is always shown to the trusted friends. However, she feels that the trust system that provides access control based on the trust value should also consider her opinion about her friends and also the system should consider other metrics such as interaction and dynamicity of the user while sharing the data. Hence, as Alice determines how much the data is sensitive, the system lists all the friends in her friend lists who are allowed to see the photo. Now, Alice's motive of sharing the data with close friends or trusted friends is satisfied.

3 Related Work

As mentioned in [5], the access control policies are designed keeping in mind the security level of the data and then the friends to whom the that data should be shown. However, the trust factor was not considered in the access control. [6] introduces the actual concept of trust in social network by calculating credibility, reliability through the peer interactions. However, more parameters can be considered to calculate trust amongst the owner and his friends. Trust Based Access Control for social networks (STBAC) was proposed which allows the trust computation amongst the owner and his friends and the data is given the access depending upon the trust between the owner and his friends. The interactions like messages and tags that the owner of the profile shares with his friend is the metric that is used to compute the trust value between the owner and friend. However, there is a disadvantage for this metric. The messages or tags cannot alone determine the trust value. Also it may happen that a good old friend may not chat on the social network much. Hence, more metrics are needed to be taken into account in order to make trust computation to be more efficient. Also according to the statistics [4] private messages are not frequently used. Hence, the interaction cannot be considered to be the sole criteria to evaluate the trust score. [7] takes into consideration the distance metric in the social network for determining the trust value and uses this trust value to filter out the content. [7] also uses clustering techniques to evaluate trust. Various techniques that were used in evaluation the trust are mentioned in [5] such as machine-based, behavioural, statistical and heuristic based techniques. [8] discusses various propagation models for trust and classification schemes for trust metrics. Also Appleseed was proposed in [8] to compute trust in local group. In [9] the distance metric is used. Hop based technique is used to decide the trust value between the users in OSNs. Lesser the hop distance more trusted the user is. Hence, in this approach the user's opinion about that particular friend is not considered. Also [12] proposes a new algorithm for trust based inference in social network. It uses probabilistic sampling and hence, calculates the trust for every source in the social network. [13] proposes finding optimal trust path between service provider and service consumer. As there are many social trust paths available, selecting the optimal one is a tough task. [13] proposes different heuristic algorithms to achieve the same. The work in [15] throws light on the pattern of usage of Facebook which tells us the popularity of the OSN sites and the extent to which they are used.

4 Evaluation of Related Work

The above related work was studied to understand the different techniques that are used to calculate the trust between the users in OSN. Each work throws light on the different methods with which the trust was calculated. The works that were studied considered trust for the purposes different than access control like for recommendation systems. Also consideration of user's opinion in the trust calculation is also vital which was not considered by many referred works. Real life trust factor (EX in our case) from the owner's perspective will always prove to be most efficient method to calculate trust.

5 Proposed Work

Considering the literature survey that is done, it is concluded that not much work has been done on the user trust in OSN. However, the essence of trust was added without considering the opinion of the user. It was observed that [11] among the users that were surveyed, many users are not much concerned about whom they are adding as friends. This is the reason why the data is leaked amongst the friends who are not intended to view that data. This leads to data confidentiality attack. The main idea behind the work is to add the flavour of trust in OSNs which depends on the system generated observations as well as the user's experience about that particular person.

Table 1 Comparison of related work [16]

Reference	Technique Used to Achieve Trust	Is Trust Used for Access Control?	User Opinion Considered?	Consideration of Characteristics of Friends in OSN
Estimating trust value: A social network perspective [7]	Clustering methods, user generated ratings	No	Yes	No
New Algorithm for Trust Inference in Social Networks [12]	Probabilistic models	No	No	No
Experimental Analysis on Access Control Using Trust Parameter for Social Network [6]	Interactions between users and friends	Yes	No	No
Propagation Models for Trust and Distrust in Social Networks [8]	Propagation models	No	No	No
Finding the Optimal Social Trust Path [13]	Heuristic algorithms	No	No	No
Multiparty Access Control for Online Social Model and Mechanisms [5]	Trust is not considered	No	No	No
Operators for Propagating Trust and their Evaluation in Social Networks [14]	Trust metrics	No	No	No
Trust based approach for protecting user data in social networks [9]	Hop based technique	Yes	No	No
Proposed Scheme	Using experience, Context Information and Interaction	Yes	Yes	Yes

5.1 Calculating Trust Score

To calculate the trust in OSN following attributes are considered:

5.1.1 Experience (EX)

Experience is the user's experience with a particular person in the real life. A person tends to accept the friend request on the social networking site but may not have good experience with him in the real life. This is not system generated but is considered as the input to the system by the user. The experience means knowing that person in real life. It was observed that [3] among the surveyed people, about 82% add only those friends who they know in real life. However, knowing a person and trusting the person are two different terms. Hence, experience plays a vital role in determining the trust amongst the owner and his friends. The system motivates the user to classify the friends in his friend list into following three categories which can be defined as follows:

5.1.1.1 *Close friends*

People who are best known and who meet frequently and have good knowledge about each other's reactions come under this category. These are the highly trusted people with whom the person can share secured data. This feeling of trust is generated from daily interactions. For example, colleagues, best friends, relatives, family members, neighbours etc. However, whom to choose into this category is solely the user's choice as the definition of 'close friends' varies from person to person.

5.1.1.2 *Friends*

These are the general friends who are not much in contact. Also, the experience with them may be average or good but not too strong to give them the access of the secured data. These are the friends that have met in the real life but not so much that they were able to build a very good trust to share the secured data. For example, the new relations that were recently built up, colleagues belonging to same institution or company met only a few times.

5.1.1.3 *Less known*

These include the friends that are known but may not have very good relations with the person. Also the person may have very good past experiences with that particular person but that may not be the same with the present time. For example, the school going friend who has met you after long time, friend of friend, person with whom very less real life experiences are shared etc. Depending on the opinion chosen by the user, the crisp values are denoted for every fuzzy value.

Table 2   Values for EX

Fuzzy Value	Crisp Value
Close Friends	1
Friends	0.75
Less known	0.25

5.1.2 Context information (CI)

This mainly includes the two sub attributes that is, number of friends the person has and the dynamicity the person has on the OSN. Dynamicity refers to how much active the user is on the OSN. The dynamicity includes the number of posts, comments, frequency of change in profile etc. According to [3] all these factors are taken into consideration by the users while they accept the friend request. Here, we consider a factor which tells the deviation factor for number of friends which is calculated as follows:

$$\alpha = \frac{nof - 130}{nof} \tag{1}$$

Where nof is number of friends and 130 is the average friends [3] the user has according to the survey done. Thus lesser the value of α less are the number of friends. Now to calculate the dynamicity of the user, we consider how much active the user is considering the number of posts, comments, number of photos uploaded etc and compare this count with the number of times he logged in. let m be the value of any activity by the user like number of posts, comments photos uploaded etc and let n be the number of times the user has logged in. Dynamicity (D) is calculated as follows:

$$D = \frac{n}{m} \tag{2}$$

Thus lesser the value of D more dynamic the user is. The fuzzy values for D are identified as follows:

Table 3   Fuzzy rules for dynamicity

Case	Fuzzy Value
D >= 0.75	Less Active
0.75 > D >= 0.5	Average Active
D < 0.5	Highly Active

Now CI can be collectively calculated as follows using the fuzzy rules.

Table 4 Fuzzy rules to evaluate CI

D	nof	CI	Fuzzy Value For CI
Less Active	<0.5	0.5	Inactive
Average Active	<0.5	1	Active
Highly Active	<0.5	1	Active
Less Active	>0.5	0.5	Inactive
Average Active	>0.5	1	Active
Highly Active	>0.5	1	Active

5.1.3 Interaction (I)

The concept of interaction is studied from [6] which is one of the aspect that is used to calculate trust. But as mentioned earlier it is not the only aspect. Interaction as mentioned in [6] is the message, likes or comments between the users. However, interaction is limited to the messages exchanged between the users. The interaction is calculated in the same way as referred in [6]. From the interaction, credibility is counted [6]. The credibility has following membership values as mentioned in [6]:

Table 5 Membership values to evaluate interaction

Credibility Value (Cr)	Membership Function Value	Fuzzy Value
$Cr >= 70$	1	Frequent
$70 > Cr > 50$	0.5	Infrequent
Otherwise	0	None

Hence, the above factors can be used to derive the Trust score between the owner of the profile and all his friends in the friend list. According to [3], the survey of 1895 users was performed and they have noted the factors that people consider while they accept the friend request. Accepting the friend request is one of the action that a user does if he trusts the person. This is the ideal situation. From [3] the above factors that is, EX, I and CI was given some weights. These weights were calculated based on the percentages that were derived for each factor from the survey. The final weighted equation for the calculating Trust score required to calculate the weights of each factor (EX, CI and I). Hence, following formula was used:

$$w = \frac{Impact\ Factor}{Total\ Impact} \tag{3}$$

Here, Impact factor denotes the importance the people have given for each factor, that is, EX, CI, and I. For example as mentioned in [3], the EX is given

82% which is the highest importance given by the people when they accept the friend request. Hence,

$$Impact\ Factor\ (EX) = 0.82 \qquad (4)$$

In the same way Impact factor is calculated for CI and I based on the real life survey in [3].

$$Total\ Impact = \sum Impact\ Factors\ (EX, CI, I) \qquad (5)$$

Based on these calculations the final Trust equation was derived. The trust equation is defined as the weighted equation where weights are derived from Equation 3. Thus, the Trust score between the owner O with the friend F_i is defined as:

$$T_{O->Fi} = 0.614EX + 0.277CI + 0.109I \qquad (6)$$

Here the variable $T_{O->Fi}$ denotes the trust score from owner O to friend F_i. Hence, if there are n friends in the friend list of the owner O there are n trust values as shown below. Consider a set T of trust scores calculated from Equation 6. Hence, T is denoted as follows for all the n friends in the friend list of O.

$$T = \{T_{O->F1}, T_{O->F2}, ... T_{O->Fn}\} \qquad (7)$$

Thus, this matches with the concept real life trust. In the real life, every person has different level of trust on the other person. This means that the trust factor changes from person to person. Hence, Equation 6 also depicts the same concept. Following cases and possible values are derived for the Trust score taking into consideration all the combinations of values of EX, CI and I. These values are derived taking into consideration all the crisp values.

From the Table 6 it is observed that minimum trust can never be 0 as even though stranger is added he will always have some or little activity on OSN. This however is the ideal case. Figure 2 shows pictorially how the trust score is calculated. Also the Figure 2 depicts how the trust score calculation is dependent on the system process and user's opinion as well. The maximum trust is 1 when owner has highest trust in the friend F_i and F_i is active on the OSN and has frequent interactions with owner. It is observed that as the majority of the users give highest importance to the real life experiences with their friends, hence, the weight for EX is highest and highly affects the trust score. From the Table 6 it is seen that even if there are no interactions with

Table 6 Trust score values

Case	EX	CI	I	Trust Score (T)
1	Close Friend	Active	Infrequent	0.943
2	Close Friend	Active	Frequent	1
3	Close Friend	Active	None	0.891
4	Friend	Active	Infrequent	0.636
5	Friend	Active	Frequent	0.688
6	Friend	Active	None	0.584
7	Less Known	Active	Infrequent	0.482
8	Less Known	Active	Frequent	0.534
9	Less Known	Active	None	0.430
10	Close Friend	Inactive	Infrequent	0.804
11	Close Friend	Inactive	Frequent	0.856
12	Close Friend	Inactive	None	0.752
13	Friend	Inactive	Infrequent	0.497
14	Friend	Inactive	Frequent	0.549
15	Friend	Inactive	None	0.445
16	Less Known	Inactive	Infrequent	0.344
17	Less Known	Inactive	Frequent	0.396
18	Less Known	Inactive	None	0.292

the best friend of owner on the OSN like Facebook and even if the friend is not active (case 12) also owns a trust 0.752. The graph in Figure 3 shows how the trust score varies with the varying values of EX, I and CI. All the 18 cases from the table are represented in the above graph. From the Figure 3 it is observed that all the values range between 0 and 1. All the trust scenarios are considered in the graph. Also, the minimum trust in the worst case is 0.292. This is the trust score when the owner doesn't have good experience with the friend F_i, and F_i is not active and there are no interactions between owner and F_i. Though it is less than many cases shown in Figure 3, but is high enough to share secured data. Hence, even if the friend is less active and has not interacted with the owner does not mean that he is not trusted at all. Hence, the user input that is EX carries more weight to calculate the trust score so that owner's best and trusted friends in any case should not be deprived of valuable information from the owner. Also, the graph shows variation of Trust score with the other varying parameters like I, CI and EX. The trust score varies according to the weighted equation that is the trust score varies linearly with the variables that is, the attributes like EX, CI and I. Thus trust score has linear variation with all the attributes used to calculate the trust score.

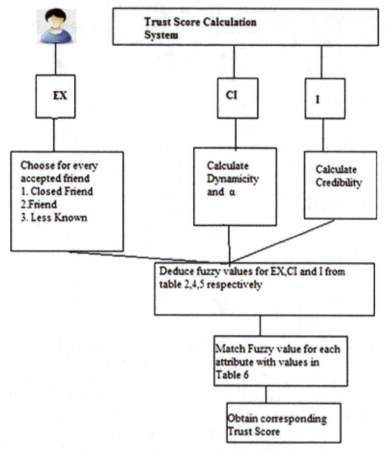

Figure 2   Flow of trust score calculation.

5.2 Trust Rule

Trust score was mainly calculated to achieve access control to the confidential or secured data over OSN. Trust Rule defines the method by which this access control can be achieved. For every data that is uploaded by the user, he has to mention the security level. Hence, with every data that is uploaded by the owner a security level [5] is attached to it. For example if Alice uploads the photo of nature's scene from the hill station that she has visited recently, that will have very less security level. Instead if she uploads the photos of her with her friends and wants that only close friends should view it, she would assign high security level to it. This security level is the input from Alice as

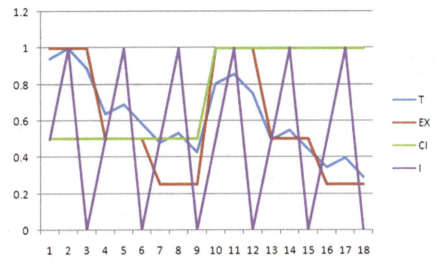

Figure 3 Graph representing the variations of trust score with variations in EX, CI and I (T).

she uploads the data. This security level is also called as trust threshold (T_{TH}). This is because:

$$Security\ Level \propto Trust\ Score \qquad (8)$$

Higher the secured data, owner wants it to be seen by only the highly trusted users. Thus the security level that is given by the owner is same as the threshold trust score (T_{TH}). Now the Trust Rule is defined as follows:

Table 7 Trust rule

Case	Access Decision
$T_{O->Fi} < T_{TH}$	Access Denied
$T_{O->Fi} >= T_{TH}$	Access Allowed

$T_{O->Fi}$ is the Trust score between owner O and friend F_i. Thus Trust Rule controls the access to the data uploaded by the user and hence, is used in access control mechanism. According to the Trust Rule, only the users that are trusted by the owner are given access to the secured data while others are denied the access.

5.3 Trust Score System

As shown in the Figure 4, there are four main components that play important roles in trust score calculation and access control.

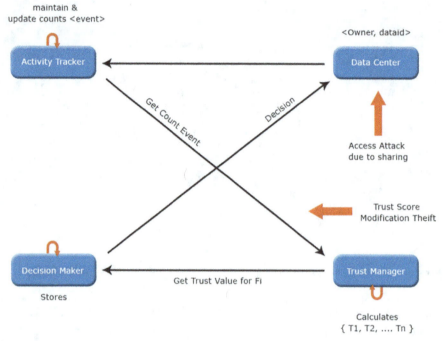

Figure 4   Trust score system.

5.3.1 Activity tracker (AT)

This component maintains and updates the required counts. For example, to calculate the dynamicity of friend of the owner, it is required to know the total number of posts and comments by the user, number of times he logged in, number of times the friend changed his profile like status, profile picture etc. All these counts are considered as the activity from the user. This activity is tracked by the Activity Tracker. The Activity Tracker stores or maintains these counts of every user along with user id. Also frequent modifications of the counts are also required. The counts are updated with the events that AT receives whenever, the user comments, posts or changes his profile details.

5.3.2 Trust manager (TM)

The main task of TM is to calculate the trust score for all n friends of the owner O. TM always receives the event from AT whenever it updates any count. With the event received from AT, TM again recalculates the trust score with the changed or updated counts and stores it in the form of $<T_{O->Fi}, O, F_i>$

Where,

$T_{O->Fi}$ is the trust score of friend F_i by the owner O.

O is the owner identified by the id

F_i is the friend id from O's friend list.

Also it recalculates the trust score if O changes EX value for F_i due to change in the real life experiences. Thus TM manages the most important job of trust score calculation. Hence, TM calculates

$$<T_{O->F1}, T_{O->F2}, ... T_{O->Fn}>$$ for all n friends in the friendlist of O.

5.3.3 Decision maker (DM)

Decision Maker is the component in the system which is responsible for the access control of the secured data. It stores all the values in the form of $<d, T_{TH}, O>$

Where,

d is the data uploaded by the owner O

T_{TH} is the security level or the threshold trust score

O is the owner.

DM takes the decision to allow or deny the access using T_{TH}, trust score from the TM and Trust Rule that was defined earlier. Thus it provides effective access control mechanism.

5.3.4 Data centre

Data centre stores all the data that is uploaded by the users of OSN. The data is stored in the 2-tuple format $<O, d>$.

Where,

O is the owner who uploads the data

d is the data id uploaded by the user

Whenever, data is uploaded on the OSN by the user, it is stored in the Data centre and its data id is generated and sent to DM along with owner id O. Data centre can be chosen to be put on the cloud. If the owner wants to change the security level of the data, it can be changed at the DM as it takes the decision regarding the decision to allow or deny access.

All the components of the system communicate with each other in the give and take of the data that is required by every component to finally achieve trust based access control.

As shown in Figure 4, two threats have been defined. Dealing with those threats is a future work.

5.4 Flow of Actions from User Perspective

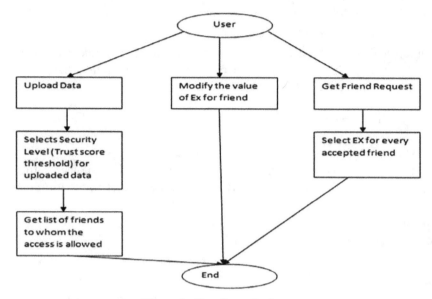

Figure 5   User flow of actions.

Figure 5 depicts the various tasks that the user performs while using the system. He uploads the data, adds the trust score threshold on which the Trust Rule gets implemented to provide the access control. Also he can view the friends to whom the access is given. With the change in the real life experience, the trust of owner towards a particular friend may increase or decrease. To absorb this real life fact in the OSN, the user can change the EX towards any of his friend. Whenever a user gets friend request he must allocate the EX to the friend. The system can be implemented using APIs as defined in [1].

6 Conclusion and Future Work

Trust plays an important role during interaction with the people in our daily life. As the people around the world are becoming more and more active on OSNs like Facebook, Twitter etc. adding the flavor of trust in OSN has also become the need of the time. Hence, an access control mechanism that works on the concept of trust was thought to be of greater importance. The proposed system calculates trust considering EX, CI and I as the three main attributes that helps in calculating the trust. Also, out of these three, CI and I are derived by the system and EX is the input from the user. The user attaches the security level with each data (photo, status, video) that she uploads. As the security level is same as available the threshold trust score, the decision is made to allow the access or deny the access depending upon the defined Trust Rule which contains the comparison between the trust score of every friend in the friend list and the threshold trust score given as the input from the user. The decision is generated by applying the Trust Rule which helps in access control of the information that is uploaded by the owner. Hence, the system achieves trust based access control using the system and user's decision which can be considered as the best and effective mechanism to calculate trust.

The threats have been identified in the system. However, to propose an effective solution against the threats is the future work. Voting mechanism can be used to handle the data confidentiality attack due to sharing of data. Also to design the algorithm for the system implementation is another future work. The system can be implemented using various APIs. (Graph API n.d.)

References

[1] Graph API available from <https://developers.facebook.com/docs/graph-api>

[2] Pewinternet http://www.pewinternet.org/2012/02/03/why-most-facebook-users-get-more-than-they-give-2/

[3] http://mashable.com/2011/12/19/friend-unfriend-facebook/

[4] https://www.stonetemple.com/how-are-people-using-facebook/

[5] Hongxin Hu, Gail-Joon, Jan Jorgensen, 'Multiparty Access Control for Online Social Model and Mechanisms', IEEE Transactions on Knowledge and Data Engineering, Vol. 25, No. 7, July 2013

[6] Saumya Omanakuttan and Madhumita Chatterjee, 'Experimental Analysis on Access Control Using Trust Parameter for Social Network', Springer-Verlag Berlin Heidelberg 2014

[7] Wei-Lun Chang & Arleen N. Diaz & Patrick C. K. Hung, 'Estimating trust value: A social network perspective', Springer Science + Business Media New York 2014

[8] Cai-Nicolas Ziegler and Georg Lausen, 'Propagation Models for Trust and Distrust in Social Networks', 2005 Springer Science + Business Media, Inc. Manufactured in The Netherlands

[9] Wilfred Villegas, 'A trust based approach for protecting user data in social networks', CASCON '07 Proceedings of the 2007 conference of the center for advanced studies on Collaborative research

[10] Chi Zhang and Jinyuan Sun, University of Florida, Xiaoyan Zhu, Xidian University Yuguang Fang, University of Florida and Xidian University. 'Privacy and Security for Online Social Networks: Challenges and Opportunities', Network, IEEE Volume: 24, Issue: 4 DOI:10.1109/MNET.2010.5510913 Publication Year: 2010

[11] Hootan Rashtian, Yazan Boshmaf, Pooya Jaferian, Konstantin Beznosov, 'To Befriend Or Not? A Model of Friend Request Acceptance on Facebook', Symposium on Usable Privacy and Security (SOUPS) 2014, July 9–11, 2014, Menlo Park, CA.

[12] Ugur Kuter, Jennifer Golbeck, 'SUNNY: A New Algorithm for Trust Inference in Social Networks Using Probabilistic Confidence Models', Copyright _c 2007, Association for the Advancement of Artificial Intelligence (www.aaai.org)

[13] Guanfeng Liu, Yan Wang, Mehmet A. Orgun, Ee-PengLim, 'Finding the Optimal Social Trust Path for the Selection of Trustworthy Service Providers in Complex Social Networks', IEEE Transactions on Services Computing, Vol. 6, No. 2, April–June 2013

[14] Chung-Wei Hang, Yonghong Wang, Munindar P. Singh, 'Operators for Propagating Trust and their Evaluation in Social Networks', 2009, International Foundation for Autonomous Agents and Multiagent Systems

[15] Duong Van Hieu, Nawaporn Wisitpongphan, and Phayung Meesad, 'Analysis of Factors which Impact Facebook Users' Attitudes and Behaviours using Decision Tree Techniques', 11th JCSSE (International Joint Conference in Computer Science and Software Engineering)

[16] Vedashree Takalkar, PN. Mahalle, 'Data confidentiality in Online Social Networks: A Survey', IJSR, Vol 4 issue 1 Jan 2015

Biographies

V. K. Takalkar graduated in Computer Science and Engineering from Pune University, Maharashtra, India in the year 2013. She has completed her Masters from Savitribai Phule Pune University. Her research interests are online network security and Internet of Things. She has published more than 6 papers in international journal and conferences. She has a teaching experience of 2 years. She is currently working as Assistant Professor in Department of Computer Engineering at Smt. Kashibai Navale College of Engineering.

P. N. Mahalle is PhD from Aalborg university and is IEEE member, ACM member, Life member ISTE and graduated in Computer Engineering from Amravati University, Maharashtra, India in 2000 and received Master in Computer Engineering from Pune University in 2007. From 2000 to 2005, was working as Assistant Professor in Vishwakarma Institute of technology, Pune, India. From August 2005, he is working as Professor and Head in Department of Computer Engineering, STES's Smt. Kashibai Navale College of Engineering, Pune, India. He published **39** research publications at national and international journals and conferences. He has authored 5 books on subjects like Data Structures, Theory of Computations and Programming Languages. He is also the recipient of "Best Faculty Award" by STES and Cognizant Technologies Solutions. He has guided more than 100 plus under-graduate students and 10 plus post-graduate students for projects. His research interests are Algorithms, IoT, Identity Management and Security.

www.ingramcontent.com/pod-product-compliance
Lightning Source LLC
LaVergne TN
LVHW012331060326
832902LV00011B/1819